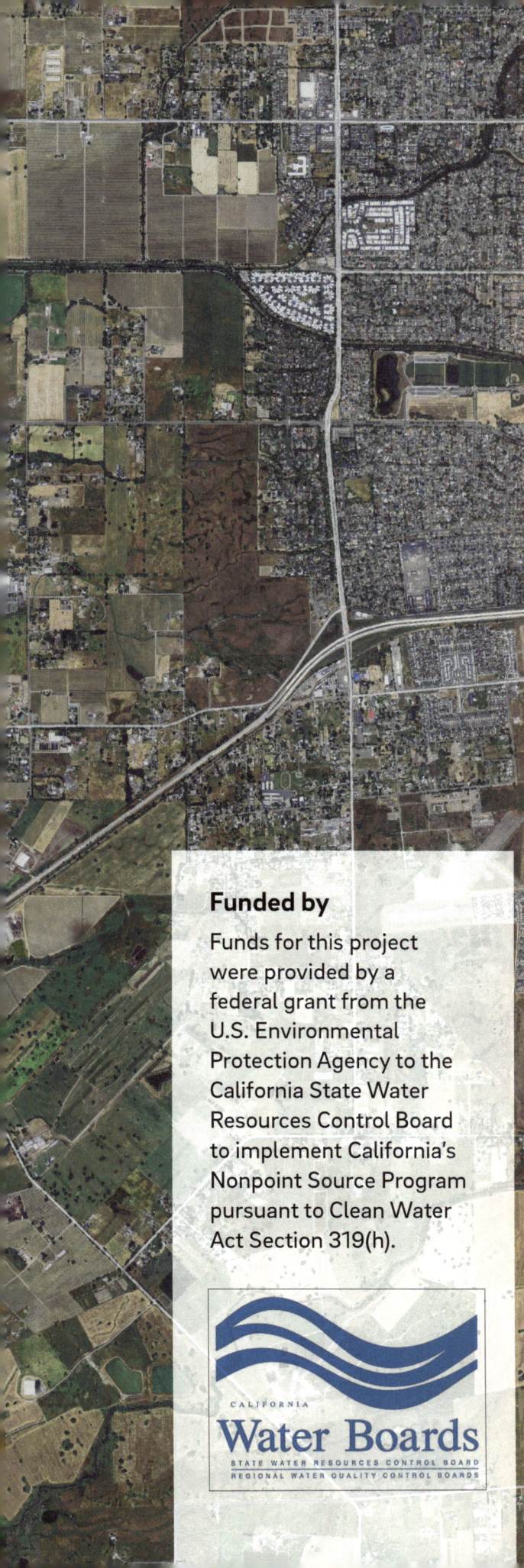

Historical Ec...
and Landscape...
in the Central

LAGUNA DE SANTA ROSA

MARCH 2017

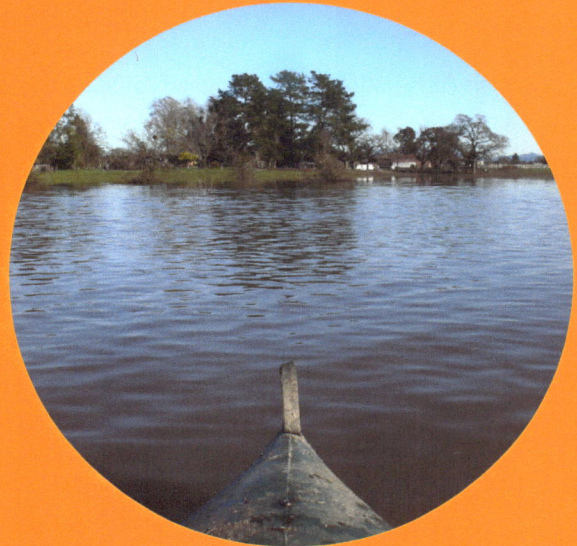

Authors

SFEI

Sean Baumgarten

Robin Grossinger

Erin Beller

Laguna de Santa Rosa Foundation

Wendy Trowbridge

DESIGN AND LAYOUT • RUTH ASKEVOLD (SFEI)

SFEI-ASC PUBLICATION #820

Funded by

Funds for this project were provided by a federal grant from the U.S. Environmental Protection Agency to the California State Water Resources Control Board to implement California's Nonpoint Source Program pursuant to Clean Water Act Section 319(h).

SUGGESTED CITATION

Baumgarten SA, Grossinger RM, Beller EE, Trowbridge W. 2017. Historical Ecology and Landscape Change in the Central Laguna de Santa Rosa. Prepared for the North Coast Regional Water Quality Control Board. A Report of SFEI-ASC's Resilient Landscapes Program and The Laguna de Santa Rosa Foundation, SFEI Publication #820, San Francisco Estuary Institute, Richmond, CA.

VERSION

March 2017 (v1.0)

REPORT AVAILABILITY

Report is available on SFEI's website at www.sfei.org/projects/LagunadeSantaRosaHE

IMAGE PERMISSION

Permissions rights for images used in this publication have been specifically acquired for one-time use in this publication only. Further use or reproduction is prohibited without express written permission from the responsible source institution. For permissions and reproductions inquiries, please contact the responsible source institution directly.

COVER CREDITS

Top: *USDC ca. 1840b, courtesy of The Bancroft Library, UC Berkeley;* Bottom: *Photo by Sean Baumgarten, January 13,2017;* Back: *see use in interior for source*

ACKNOWLEDGEMENTS

This project was funded by a federal grant from the U.S. Environmental Protection Agency to the State Water Resources Control Board to implement California's Nonpoint Source Program pursuant to Clean Water Act Section 319(h). We would like to thank David Kuszmar, Michele Fortner, Chuck Striplen, Steve Butkus, and Clayton Creager at the North Coast Regional Water Quality Control Board for their guidance, support, and feedback throughout the course of the project.

The report benefited greatly from the input of our advisory team (see page 13). Technical advisors Hattie Brown, Steve Butkus, Keenan Foster, William Hart, Kara Heckert, Conrad Jones, David Kuszmar, Eric Larson, Gaye Lebaron, Jane Nielson, Tom Origer, Geoffrey Skinner, Chuck Striplen, and Mark Tukman provided insightful guidance and comments on project analyses and reporting.

We are indebted to the staff and volunteers at the local, regional, county, and state archives that we visited over the course of the project (see page 7). In particular, we would like to thank Lynn Prime of the Sonoma State University Library; Katherine Rinehart and Tony Hoskins of the Sonoma County History & Genealogy Library; Brian Curtis and Bob Curtis of Curtis & Associates, Inc.; Sally Morrison of the Western Sonoma County Historical Society; David Conklin of the Bureau of Land Management; and Gary O'Connor of the Sonoma County Surveyor's office for their extraordinary help and patience.

We would like to thank SFEI staff, including Julie Beagle, Emily Clark, Josh Collins, Scott Dusterhoff, Letitia Grenier, Steve Hagerty, Micha Salomon, David Senn, and Jing Wu for their assistance with data collection, GIS mapping, analyses, graphics development, and report development. Jay Scherf, an intern from UC Berkeley, contributed greatly to data collection efforts. Thank you to Kate Lundquist at the Occidental Arts and Ecology Center for her helpful insights and responses to questions.

We are very grateful to the staff and volunteers at the Laguna de Santa Rosa Foundation for all of their support and assistance throughout the course of the project. Hattie Brown (Sonoma County Regional Parks, formerly Laguna de Santa Rosa Foundation) deserves many thanks for all of her contributions to the project.

CONTENTS

INTRODUCTION

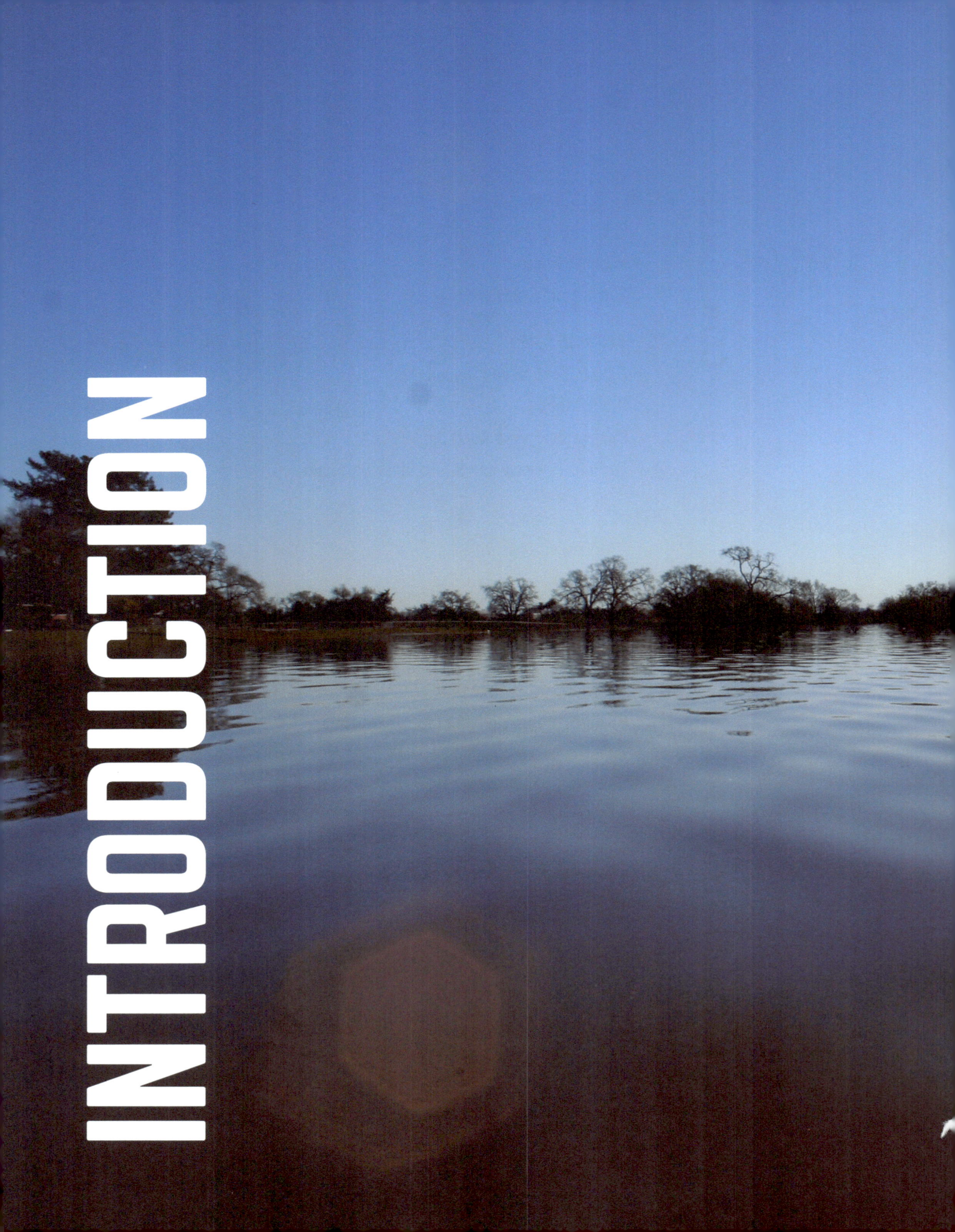

This study synthesizes a diverse array of data to examine the ecological patterns, ecosystem functions, and hydrology that characterized a central portion of the Laguna de Santa Rosa during the mid-19th century, and to analyze landscape changes over the past 150 years. The primary purpose of this study was to help guide restoration actions and other measures aimed at reducing nutrient loads within this portion of the Laguna de Santa Rosa watershed.

Prior to Euro-American modification of the landscape, the Laguna de Santa Rosa (hereafter, "Laguna") supported a diverse and extensive array of wetland and riparian habitat types ranging from tule marshes and willow forests to wet meadows and vernal pools. As one of the largest wetland complexes in northern California, the Laguna was a productive ecosystem that was home to vast numbers of resident and migratory birds, fish, mammals, and other wildlife. It was this diversity and productivity that made the Laguna a focal element of the Santa Rosa Plain for generations of native Pomo people, and for 19th-century Spanish and American settlers.

Today, the Laguna continues to provide vital habitat for a host of native plants and animals, and has been recognized by the Ramsar Convention on Wetlands as a Wetland of International Importance (Ramsar 2017). However, as the population of the area has grown and land use modifications have intensified over the past 150 years, the Laguna has experienced a variety of impacts stemming from urban and agricultural development. These impacts include substantial loss of native wetland and riparian habitats, channelization of streams, alterations to streamflow and sediment dynamics, invasion of non-native species, and water quality impairments. While the Laguna still exhibits elements of its former character, the modern Laguna represents a landscape profoundly altered by over a century and a half of both direct and indirect modifications. As residents and land managers look to improve the ecological health of the Laguna and restore lost ecological functions, an understanding of what the landscape was like in the recent past, and how it has changed over time, provides a valuable starting place for envisioning its future potential.

PROJECT BACKGROUND AND OBJECTIVES

This project, funded by the State Water Resources Control Board, developed historical information for a central portion of the Laguna de Santa Rosa (see page 4 for study area). As one component of an ongoing initiative to reconstruct historical landscape patterns and processes within the Laguna watershed, this study built on

Photo by Sean Baumgarten, January 13, 2017

several previous historical research efforts, including a study of changes in channel alignment along the lower Laguna de Santa Rosa and Mark West Creek (Baumgarten et al. 2014a) funded by the Sonoma County Water Agency, and the development of historical orthoimagery for the Santa Rosa Plain (Baumgarten et al. 2014c) funded by the Sonoma County Agricultural Preservation and Open Space District. In addition, this study drew on an extensive body of existing contemporary research, planning efforts, and technical documents, including *Enhancing and Caring for the Laguna* (Honton and Sears 2006) and *The Altered Laguna: A Conceptual Model for Watershed Stewardship* (Sloop et al. 2007). Future research in collaboration with the Sonoma County Water Agency and the Laguna de Santa Rosa Foundation, funded by the California Department of Fish and Wildlife, will expand the historical research presented here and use it to inform the development of a restoration plan for the Laguna (see page 55).

The Laguna has long experienced water quality challenges, including excessive nutrient concentrations. In the late 1800s and early 1900s, raw sewage was discharged into the Laguna and Santa Rosa Creek from the nearby towns of Santa Rosa and Sebastopol (*Daily Alta California* 1888, Searls 1897, Lee 1944, Tancreto and Rivera 1987, Cummings 2003b). During the mid-20th century, several wastewater treatment plants (including the Laguna Treatment Plant) were constructed to address this problem, which (thanks in part to upgrades in recent decades) have greatly reduced nutrient concentrations in discharged wastewater (Tancreto and Rivera 1987, Sloop et al. 2007). Today, one of the largest sources of ongoing nutrient inputs to the Laguna is the large number of dairies and other agricultural enterprises located within the watershed. In addition, legacy discharges stemming from sewage, industrial tailings, manure, and other sources continue to be a significant contributor to overall nutrient loads in the Laguna (TetraTech 2015a, D. Kuszmar pers. comm.). The increased rate of nutrient loading over historical levels is compounded by the substantial alterations to the Laguna's drainage network (e.g., straightening and channelization of streams) and the loss of a wide variety of wetland types in the watershed, which have significantly reduced nutrient assimilative capacity (the ability of wetlands to remove or filter nutrient loads) and impacted flow and sediment dynamics. Biostimulatory conditions associated with these changes have led to eutrophication and depressed dissolved oxygen levels within aquatic and wetland habitats, contributing to fish kills, invasion of non-native aquatic plants, and other impacts (PWA 2004a; Sloop et a. 2007; Fitzgerald 2013; TetraTech 2015a, 2015b).

The Laguna was initially listed on the 303(d) list of impaired water bodies in 1990 for high ammonia and low dissolved oxygen (TetraTech 2015a). A "waste reduction" strategy was developed in 1995 to address these impairments, focusing on measures that should be taken to reduce pollutant loads (Morris 1995). The Laguna was subsequently listed for sediment, nitrogen, phosphorus, dissolved oxygen, temperature, mercury, and indicator bacteria (Fitzgerald 2013). The North Coast Regional Water Quality Control Board is currently developing Total Maximum Daily Loads to address these impairments and identify source reduction targets to restore beneficial uses outlined in the Water Quality Control Plan for the North Coast Region; these beneficial uses include recreation, water supply, groundwater recharge, freshwater habitat, biodiversity, and flood storage/attenuation, among others (NCRWQCB 2011). However, pollutant load reduction is only one part of a comprehensive strategy to address excess nutrient loading: strategic restoration of the Laguna's wetland and riparian buffer can help to increase the assimilative capacity of the system, and thus reduce nutrient loading to surface waters. Ultimately, both source controls and strategic habitat restoration will be needed to address water quality problems and maximize ecological functions within the Laguna wetland complex.

Determining what habitat types should be restored, and in which areas, requires a thorough understanding of the ecological patterns that existed historically, the natural processes that sustained them, and the potential for enhancing or reestablishing those processes. The study of the past can help inform restoration decisions by providing valuable knowledge about system characteristics under more natural conditions, as well as an understanding of how these characteristics have changed over time in response to human alterations to the landscape. Understanding the interaction between the ecological mosaic and the underlying topographic, climatic, and hydrologic gradients is key to designing and managing locally appropriate future systems that are flexible, adaptive, and resilient to dynamic environmental conditions.

The objectives of this study were to:

- Create a synthetic map representing average ecological conditions in the central portion of the Laguna ca. 1850.

- Link ecological conditions to physical processes and environmental drivers, and document how these conditions and drivers have changed over time.

- Enhance understanding of how changes in wetland extent have impacted nutrient dynamics and assimilative capacity.

- Foster a shared understanding of local landscape history and habitat values, establishing a common reference point across diverse stakeholders.

REPORT STRUCTURE

This chapter provides an overview of the project background and objectives and the environmental setting of the Laguna watershed. Chapter 2 discusses the methodology used to reconstruct historical landscape patterns within the study area and to analyze landscape changes over time. Chapter 3 presents research findings related to historical ecological patterns and processes, including information about the historical channel network, hydrology, habitat types, and biological diversity. Chapter 4 summarizes land-use changes within the watershed, analyzes their impacts on habitat extent/distribution and channel configuration, and explores the effects of landscape changes on nutrient assimilative capacity. The report concludes with a discussion of management implications, restoration potential, and future research directions in Chapter 5. Scientific names of species referenced in the report are provided in the Appendix.

A LAGUNA DE SANTA ROSA WATERSHED

Sonoma County

Napa County

Russian River

Windsor

Mark West Creek

Forestville

S116

Santa Rosa Creek

Santa Rosa

US101

Sebastopol

Laguna de Santa Rosa

Rohnert Park

Marin County

S1

N

2 miles

B PROJECT STUDY AREA

Santa Rosa Creek

Laguna de Santa Rosa

Irwin Creek

Occidental Rd

Gravenstein Rd N

CA-12

Duer Creek

Lake Jonive

Gravenstein Creek

SEBASTOPOL

5 miles

SAN FRANCISCO BAY

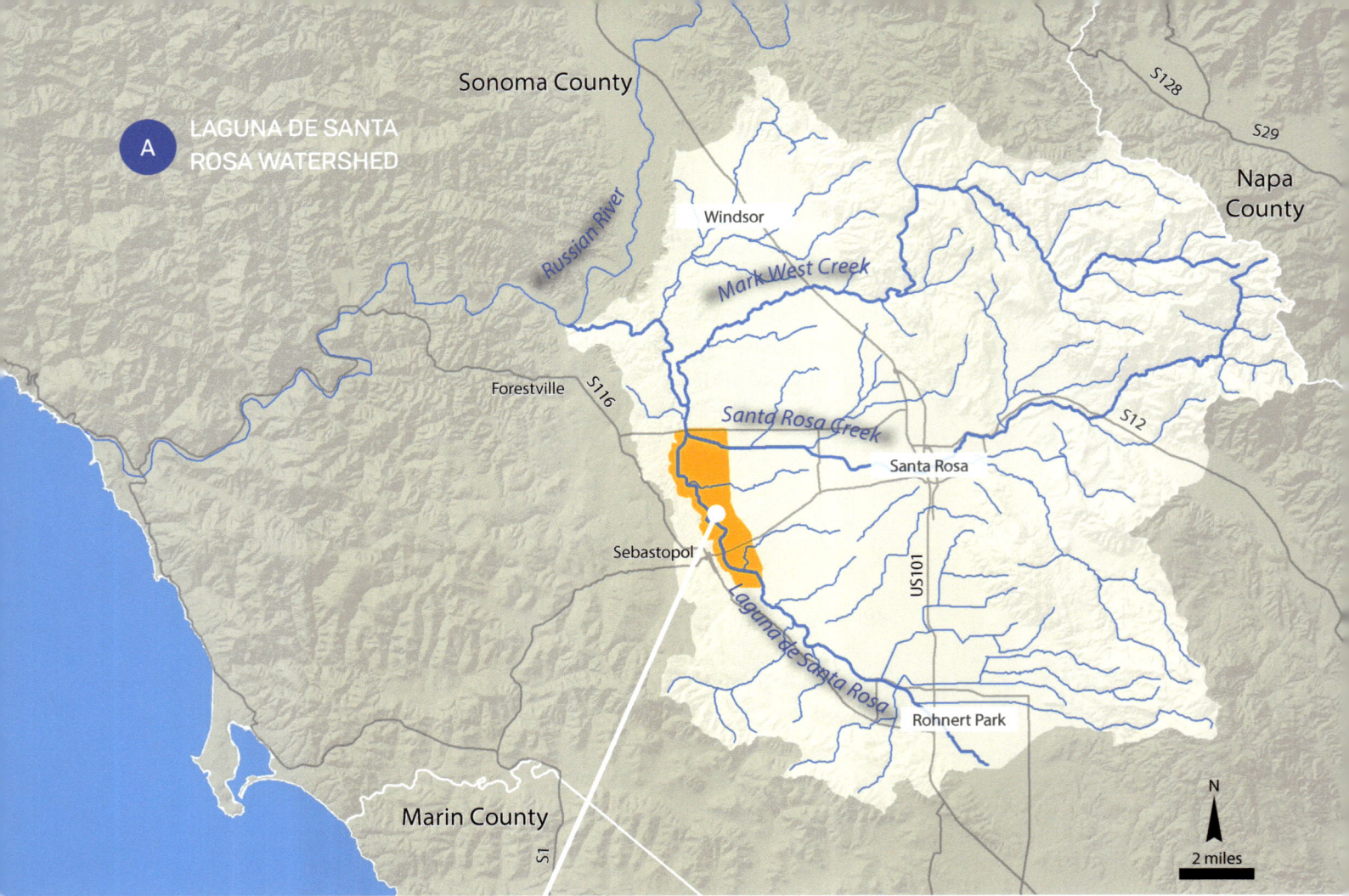

Fig. 1. The Laguna de Santa Rosa watershed (A) and project study area (B) are situated in Northern California. The study area, highlighted in orange, encompasses approximately 13 km² (5 mi²) of a central portion of the central Laguna de Santa Rosa and surrounding areas.

SETTING AND STUDY AREA

The Laguna de Santa Rosa lies at the western edge of a 658 km^2 (254 mi^2) watershed in Sonoma County, California (Fig. 1A). Numerous tributaries, including Mark West Creek, Santa Rosa Creek, Matanzas Creek, and Windsor Creek, originate in the Mayacamas and Sonoma mountains to the east and flow west across the Santa Rosa Plain. The Laguna drains into the Russian River, from which point the combined waters flow 39 km (24 mi) to the Pacific Ocean.

The Laguna is located in a tectonic depression along the boundary of two crustal blocks (Curtis et al. 2013). The Wilson Grove Formation, dominated by marine sandstone, forms the hills to the west of the Laguna (Delattre and Koehler 2008). The Sonoma Volcanics form the mountains on the eastern side of the watershed (Honton and Sears 2006). The Santa Rosa Plain, which extends from the Laguna to the Mayacamas Mountains on the east, is comprised of eroded sediments carried down from the surrounding ranges by alluvial processes.

Plant communities on the Santa Rosa Plain include grasslands, oak woodlands, and vernal pools. Interior mixed hardwood and coniferous forest dominate the hills to the east of the Santa Rosa Plain (Honton and Sears 2006, USDA 2009). Altogether, open space (including rangeland) comprises approximately 14% of the watershed. Developed areas, including the cities of Santa Rosa, Rohnert Park, Windsor, Cotati, and Sebastopol, occupy approximately 53% of the watershed, while agricultural lands, concentrated on the western side of the Santa Rosa Plain, occupy 33% (ABAG 2005).

The study area encompasses approximately 13 km^2 (5 mi^2) in the central portion of the Laguna de Santa Rosa to the east of Sebastopol (Fig. 1B). It extends from Guerneville Road on the north to approximately 7 km (4.5 mi) south. The western boundary of the study area is approximately defined by the geologic divide between the alluvium of the Santa Rosa Plain and the bedrock of the Wilson Grove Formation. The eastern boundary encompasses all of the wetland features contiguous to the Laguna, with the exception of oak savanna/vernal pool complex and wetlands extending upstream along the Laguna mainstem and Santa Rosa Creek.

METHODOLOGY

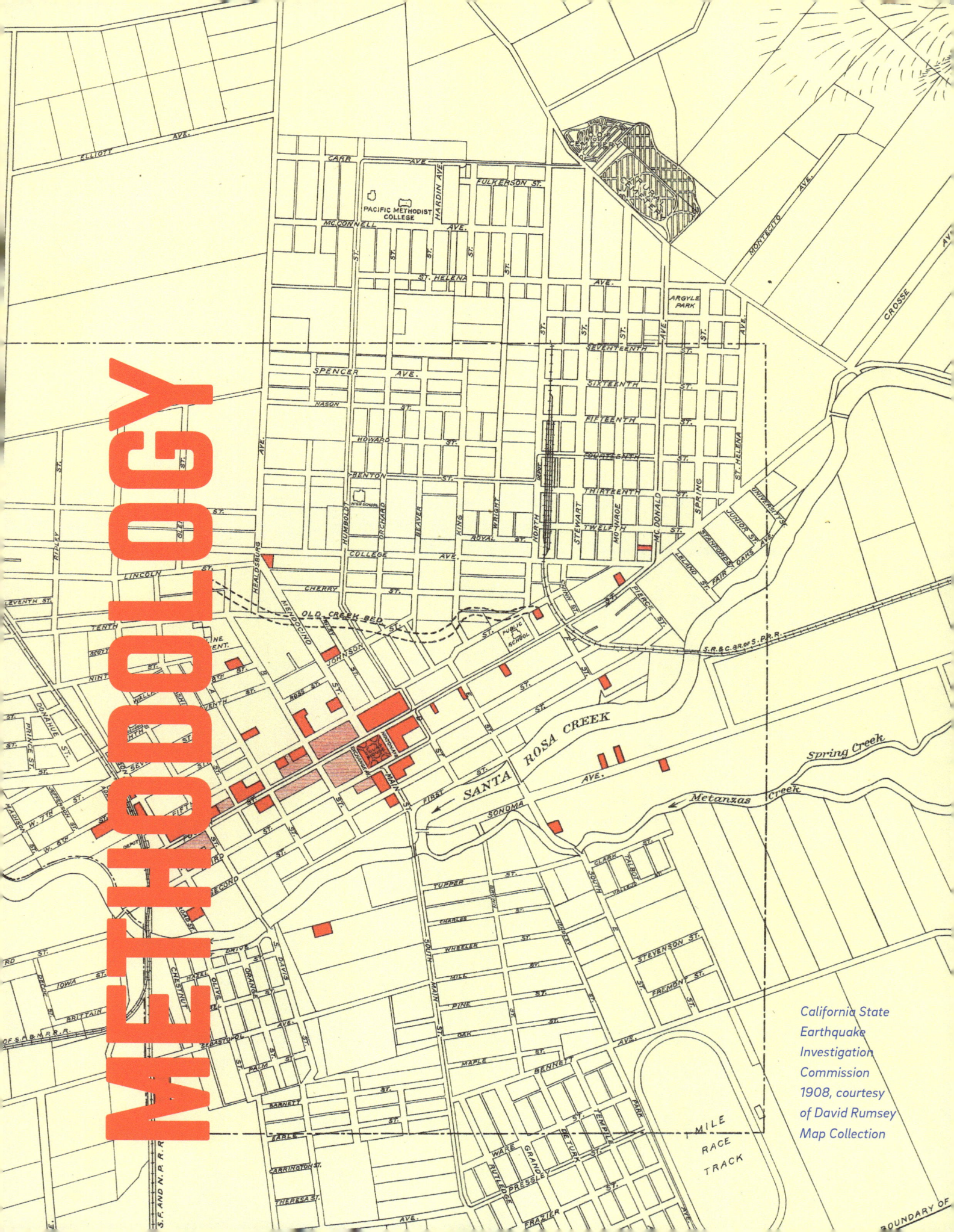

California State Earthquake Investigation Commission 1908, courtesy of David Rumsey Map Collection

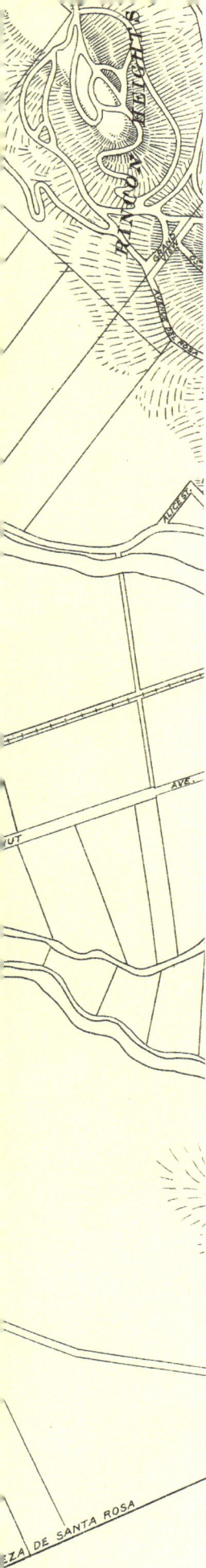

The use of historical data to study past ecosystem characteristics is an interdisciplinary field called historical ecology (Swetnam et al. 1999, Rhemtulla and Mladenhoff 2007). Historical ecology can reveal relevant landscape drivers and context, document change over time, and suggest appropriate restoration and management targets by identifying constraints and opportunities posed by the contemporary landscape.

Constructing an accurate picture of historical landscape patterns requires the integration, comparison, and interpretation of many independent sources (Grossinger et al. 2007). Where possible, historical landscape features in the Laguna were documented using multiple sources from varying years and authors to ensure accurate interpretation. This section details how these sources were collected and interpreted, as well as how they were used to create maps of historical habitat types and channels.

DATA COLLECTION AND COMPILATION

Archival data used to reconstruct the historical Laguna landscape were collected from 13 local, regional, county, and state archives (Table 1), as well as over 20 online databases. The dataset includes a substantial amount of information collected for previous historical ecology research in the Laguna, Dry Creek, and Russian River watersheds (e.g., Baumgarten et al. 2014a, 2014b, 2014c) and ongoing research in the Petaluma River watershed. Data collection efforts focused on information related to hydrology, channel configuration, and habitat extent, distribution, and composition; information about other topics, such as wildlife presence or botanical diversity, was collected where relevant but was not the target of data collection efforts.

The assembled dataset is composed of a variety of historical data types, including maps (e.g., Spanish land grant maps, General Land Office [GLO] survey plats, USGS and USACE topographic quads, USDA soil maps, county maps, parcel and subdivision maps, railroad maps), photographs (landscape and aerial), and textual documents (e.g., land grant case files, GLO field notes, travelogues, newspaper articles, county histories, specimen records). It also includes contemporary technical reports and Geographic Information Systems (GIS) data layers such as the Sonoma County LiDAR (WSI 2013), SSURGO soil data, and modern aerial photos. Altogether, the dataset includes approximately 500 landscape and oblique aerial photographs, 200 maps, and 150 textual documents (Fig. 2).

Table 1. Source institutions visited for project data collection.

SOURCE INSTITUTION	LOCATION
Bureau of Land Management	Sacramento (visited remotely)
California Historical Society	San Francisco
Curtis & Associates, Inc.	Healdsburg
North Coast Regional Water Quality Control Board	Santa Rosa
Sonoma County Public Library and History & Genealogy Annex	Santa Rosa
Sonoma County Recorder	Santa Rosa
Sonoma County Surveyor	Santa Rosa
Sonoma County Water Agency	Santa Rosa
Sonoma State University	Rohnert Park
The Bancroft Library, UC Berkeley	Berkeley
The George & Mary Foster Anthropology Library, UC Berkeley	Berkeley
UC Berkeley Earth Science and Map Library	Berkeley
Western Sonoma County Historical Society	Sebastopol (visited remotely)

Sources with high spatial accuracy and/or detailed depictions of landscape features were compiled and georeferenced in a GIS database. The georeferenced data includes approximately 25 historical maps, as well as spatially explicit narrative data such as GLO field notes, county survey notes, and records from species databases such as the Consortium of California Herbaria (CCH) and the California Natural Diversity Database (CNDDB).

One key data source for the study was a historical aerial photomosaic of the entire Santa Rosa Plain (Fig. 3) comprised of orthorectified and mosaicked aerial photos from 1942 (the earliest year for which coverage of the region is available). The photomosaic was created by SFEI in 2014 with funding from the Sonoma County Agricultural Preservation and Open Space District, and is viewable on the Sonoma Veg Map website (http://www.sonomavegmap.org/1942/).

Fig. 2. Examples of archival data sources collected for this study. Each source provides a distinct perspective on the Laguna's historical ecology. (Map: Bowers 1866, courtesy of David Rumsey Map Collection; Photo: CHS-45682, courtesy of USC Libraries, California Historical Society Collection; Newspaper: Daily Alta California 1872, courtesy of California Digital Newspaper Collection)

Fig. 3. (opposite) Photomosaic of historical aerial imagery, flown in 1942, of the Santa Rosa Plain. (USDA 1942)

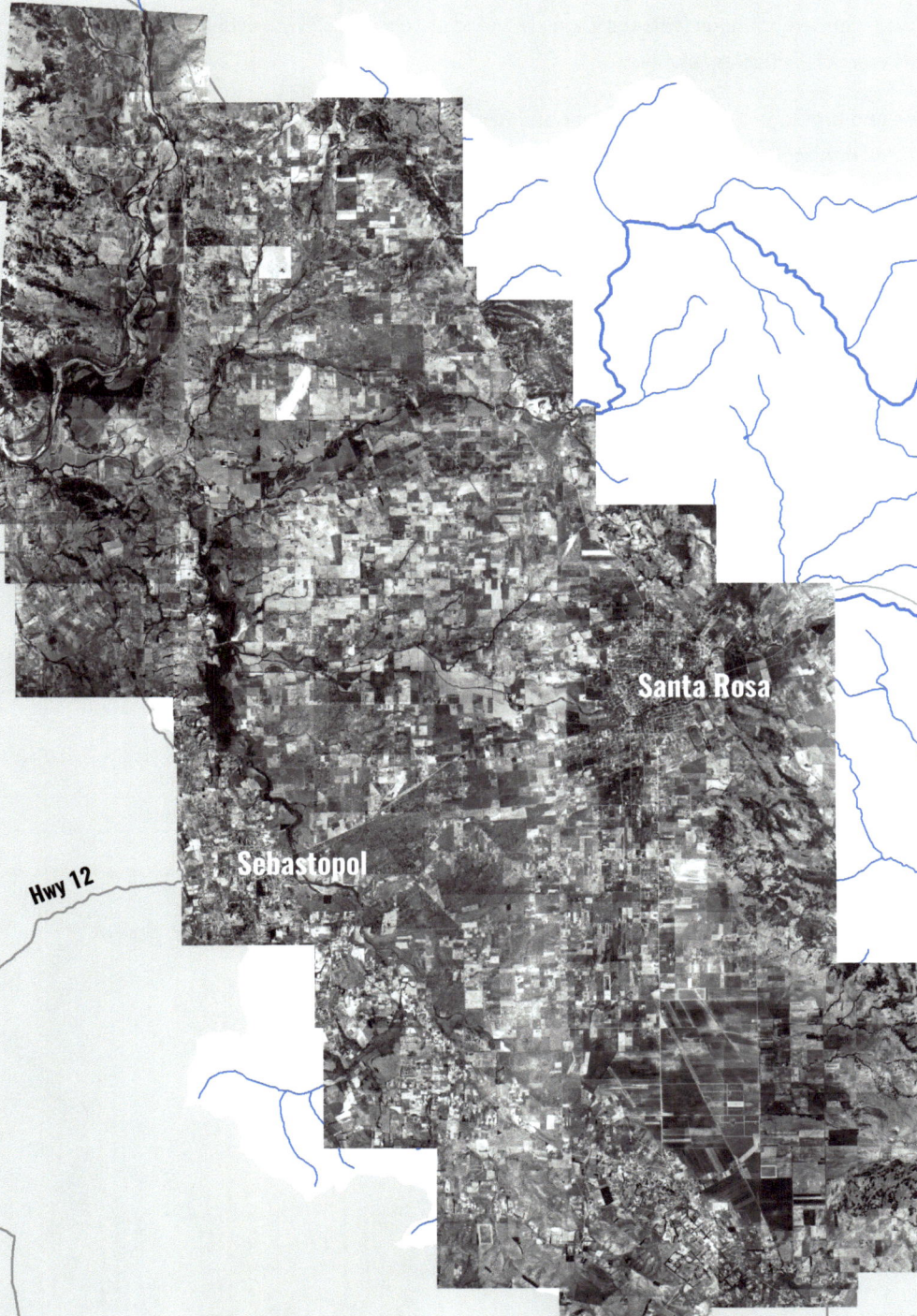

Hwy 128

Hwy 29

Santa Rosa

Sebastopol

Hwy 12

Hwy 12

Hwy 101

N

2 miles

GIS SYNTHESIS

The collected data were synthesized and interpreted to develop two GIS layers representing the historical distribution of wetland, riparian, and aquatic habitat types and the configuration of the drainage network during the mid-19th century. Habitat types were mapped as polygon features (Table 2). Drainage features, including the Laguna mainstem and significant tributaries, were mapped as line features. Streams and other relatively well-defined drainage features were classified as "channels," while shallower, low-gradient drainage features were classified as "sloughs."

Habitat polygons and drainage features were mapped from the most spatially accurate sources believed to be representative of historical landscape condition and configuration. In many cases, features were digitized from either the 1942 aerial photographs (USDA 1942) or the modern LiDAR-derived DEM (WSI 2013), while in other cases features were digitized from spatially accurate historical maps such as GLO plats. Multiple early sources (i.e., mid-late 19th-century maps and textual data) were used to confirm the historical presence of particular features and establish their approximate shape, size, location, and classification. Mapped features were attributed in the GIS with supporting sources as well as certainty levels for shape, location, and interpretation (classification).

Table 2. **Wetland, riparian, and aquatic habitat types** captured by the historical synthesis map, along with characteristic vegetation components and flood regime.

HABITAT TYPE	CHARACTERISTIC VEGETATION COMPONENTS	FLOOD REGIME
Perennial Freshwater Lake/Pond	None, floating	Perennial
Valley Freshwater Marsh	Tule, cattail	Semi-Permanent to Perennial
Willow Forested Wetland	Willow, tule, Oregon ash	Seasonal to Semi-Permanent
Mixed Riparian Forest	Oak, willow, Oregon ash, boxelder	Temporary
Wet Meadow	Grasses, rushes, forbs	Temporary to Seasonal
Oak Savanna/Vernal Pool Complex	Vernal pool specialists, oaks, dry grasses	Temporary to Seasonal

Perennial Freshwater Lake/Pond

Valley Freshwater Marsh

Willow Forested Wetland

Wet Meadow

Mixed Riparian Forest

Oak Savanna/Vernal Pool Complex

Perennial Freshwater Lake/Pond: Photo by Micha Salomon, June 6, 2013; Valley Freshwater Marsh: Photo by Jean Pawek, 2013, courtesy of Calflora; Willow Forested Wetland: Photo by Brian Bollman, January 5, 2013, https://flic.kr/p/dMoDNw; Wet Meadow: Photo by Stickpen, October 2009, courtesy of Wikimedia Commons; Mixed Riparian Forest: Photo by Micha Salomon, June 6, 2013; Oak Savanna/Vernal Pool Complex: Photo by Josh Collins

CHANGE OVER TIME ANALYSIS

A map of contemporary channels and wetland, riparian, and aquatic habitat types (ca. 2015) was developed in order to analyze changes in habitat distribution and channel configuration over time. The contemporary map was based primarily on North Coast Aquatic Resource Inventory (NCARI) mapping (SFEI-ASC 2014). Additional riparian features (classified as "Southwestern North American Riparian Evergreen and Deciduous Woodland Group" or "Southwestern North American Riparian/Wash Scrub Group") were incorporated from a draft version of the Sonoma Veg Map GIS layers (http://sonomavegmap.org/). Modern classifications were crosswalked to historical habitat types to enable comparison between the historical and modern mapping (Table 3).

Table 3. Crosswalk between modern classifications and historical habitat types. Modern wetland classes follow the classification systems used by NCARI (SFEI-ASC 2014) and a draft version of the Sonoma Veg Map (http://sonomavegmap.org/).

SOURCE	MODERN CLASSIFICATION	ABBREV.	HISTORICAL ANALOG
NCARI	Channel Open Water Natural	COWN	1) Lake Jonive and similar large open water features → Perennial Freshwater Lake/Pond 2) For narrow channels, classification was merged with adjacent polygons
NCARI	Channel Vegetated Natural[1]	CVN	Valley Freshwater Marsh
NCARI	Depressional Open Water Natural	DOWN	Perennial Freshwater Lake/Pond
NCARI	Depressional Open Water Unnatural	DOWU	1) For large polygons in the northern part of the study area with aquatic vegetation visible in NAIP 2012 → Open Water/Aquatic Vegetation 2) For managed ponds or other small, unnatural open water features outside the core of the Laguna → Storage Pond 3) For small open water features embedded within other wetland types in the core of the Laguna → Perennial Freshwater Lake/Pond 4) Several polygons were classified as Open Water/Agriculture based on local expert knowledge
NCARI	Depressional Vegetated Natural	DVN	Valley Freshwater Marsh
NCARI	Depressional Vegetated Unnatural	DVU	1) Default = Valley Freshwater Marsh 2) One polygon was classified as Open Water/Agriculture based on local expert knowledge
NCARI	Forested Slope	FS	Forested Wetland and Riparian Forest/Scrub
NCARI	Riparian - Forested Slope	FSr	Forested Wetland and Riparian Forest/Scrub
NCARI	Lacustrine Open Water Unnatural	LOWU	Storage Pond
NCARI	Lacustrine Vegetated Unnatural	LVU	Valley Freshwater Marsh
NCARI	Natural Slope	SN	Wet Meadow
NCARI	Unnatural Slope	SU	Wet Meadow
NCARI	Individual Vernal Pool	VP	N/A
NCARI	Vernal Pool Complex	VPC	N/A
NCARI	Wet Meadow	WM	Wet Meadow
Sonoma Veg Map	Southwestern North American Riparian Evergreen and Deciduous Woodland Group		Forested Wetland and Riparian Forest/Scrub
Sonoma Veg Map	Southwestern North American Riparian/Wash Scrub Group		Forested Wetland and Riparian Forest/Scrub

[1] Within the study area, the NCARI mapping generally shows modified sections of channels as line features (rather than polygons), and thus the crosswalk does not include a category for Channel Vegetated Unnatural.

Three novel habitat types not present historically were included in the contemporary mapping: Storage Pond, Open Water/Aquatic Vegetation (areas that are often shallowly flooded and have a cover of *Ludwigia* sp. and other non-native aquatic plants) and Open Water/Agriculture (areas that are periodically cultivated but are often shallowly flooded).

Several adjustments to the historical mapping were made to facilitate comparison with modern data. First, because Oak Savanna/Vernal Pool Complex includes substantial non-wetland areas and is not directly comparable with modern vegetation classifications, it was omitted from the change analysis. Second, Willow Forested Wetland and Mixed Riparian Forest were combined into one habitat type called Forested Wetland and Riparian Forest/Scrub.

TECHNICAL REVIEW AND STAKEHOLDER OUTREACH

A technical advisory team consisting of 14 local experts and resource managers (Table 4) provided guidance on project direction, methodology, mapping, and reporting. In addition, two stakeholder workshops were held during the course of the project, including a workshop at the Laguna de Santa Rosa Foundation in October 2016 to present preliminary findings and solicit comments and feedback from local landowners and other stakeholders, and a workshop in February 2017 to present project findings to stakeholders.

Table 4. Project Technical Advisory Team. Acroymns: NCRWQCB (North Coast Regional Water Quality Control Board), CDFW (California Department of Fish and Wildlife), SCWA (Sonoma County Water Agency).

ADVISOR	AFFILIATION
Hattie Brown	*Sonoma County Regional Parks (previously Laguna de Santa Rosa Foundation)*
Steve Butkus	*NCRWQCB*
Keenan Foster	*SCWA*
William Hart	*Gold Ridge Resource Conservation District*
Kara Heckert	*Sonoma Resource Conservation District*
Conrad Jones	*CDFW*
David Kuszmar	*NCRWQCB*
Eric Larson	*CDFW*
Gaye Lebaron	*Historian*
Jane Nielson	*Geologist*
Tom Origer	*Tom Origer & Associates*
Geoffrey Skinner	*Sonoma County Library*
Chuck Striplen	*NCRWQCB (previously SFEI)*
Mark Tukman	*Tukman Geospatial*

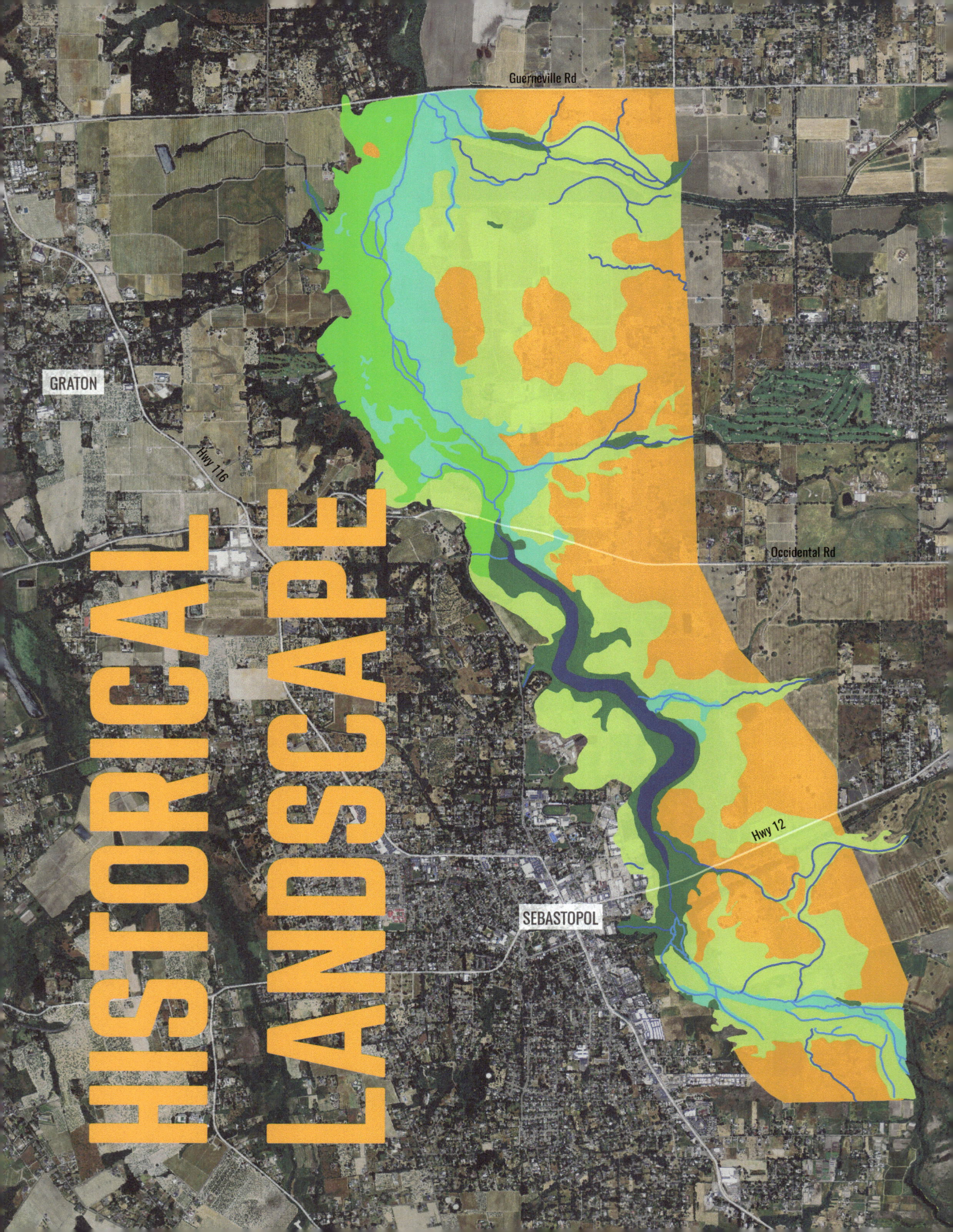

HISTORICAL LANDSCAPE

GRATON

Hwy 116

Guerneville Rd

Occidental Rd

Hwy 12

SEBASTOPOL

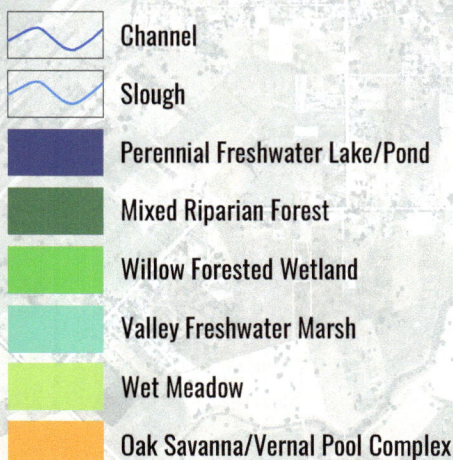

SANTA ROSA

The historical (ca. 1850) distribution of habitat types in the study area, including freshwater ponds and wetlands, riparian areas, and stream channels.

- Channel
- Slough
- Perennial Freshwater Lake/Pond
- Mixed Riparian Forest
- Willow Forested Wetland
- Valley Freshwater Marsh
- Wet Meadow
- Oak Savanna/Vernal Pool Complex

N

1/2 mile

OVERVIEW

During the mid-19th century, the central portion of the Laguna and surrounding areas (the focus of this study) were characterized by a diversity of wetland types with a complex spatial distribution. Open water bodies, perennial wetlands, and riparian habitat types formed the core of the Laguna wetland complex, while extensive seasonal wetlands and oak savannas occupied the surrounding areas, resulting in both lateral (i.e., across valley) and longitudinal (i.e., along valley) variability in wetland distribution (Fig. 4).

Entering the study area from the southeast, the Laguna flowed through a relatively narrow corridor of valley freshwater marsh surrounded by extensive areas of wet meadow and oak savanna/ vernal pool complex. As the Laguna turned north and flowed past present-day Sebastopol, the surrounding marsh graded into a mixed riparian forest dominated by willows and oaks. Just north of present-day Highway 12, the Laguna emptied into a deep, perennial open water body known as Lake Jonive, which extended north to present-day Occidental Road and was bordered by mixed riparian forest. North of Occidental Road, the width of the Laguna increased considerably: a broad swath of valley freshwater marsh extended along the eastern side of the channel, while a large willow forested wetland dominated the western side. Extensive areas of wet meadow and oak savanna/vernal pool complex bordered the perennial wetlands and riparian forests on the eastern side of the Laguna. Overall, wet meadow and oak savanna/vernal pool complex occupied the largest areas within the study area (~470 and 420 ha [1,170 and 1,050 ac], respectively), followed by valley freshwater marsh (~130 ha [310 ac]), willow forested wetland (~120 ha [290 ac]), mixed riparian forest (~70 ha [180 ac]), and perennial freshwater lake/pond (~30 ha [70 ac]; Fig. 5).

The gradual slope of the Laguna (on average, just a few vertical feet per mile) was one characteristic that fundamentally influenced hydrologic patterns and wetland distribution. The gradual slope caused water to move very slowly through the system and spread out over large wetland complexes (De Mars et al. 1977). The distribution of wetland and riparian habitat types within the study area was influenced by a number of interrelated physical controls, including topography, soils, geology, streamflow, and groundwater. Patterns of alluvial deposition along tributary streams gave rise to some of the topographic complexity and soil heterogeneity within the Laguna, with recent (Holocene) alluvial deposits created by Santa Rosa Creek and other tributaries inset within the older (Pleistocene) deposits that occupied much of the Santa Rosa Plain (Honton and Sears 2006, Delattre and Koehler

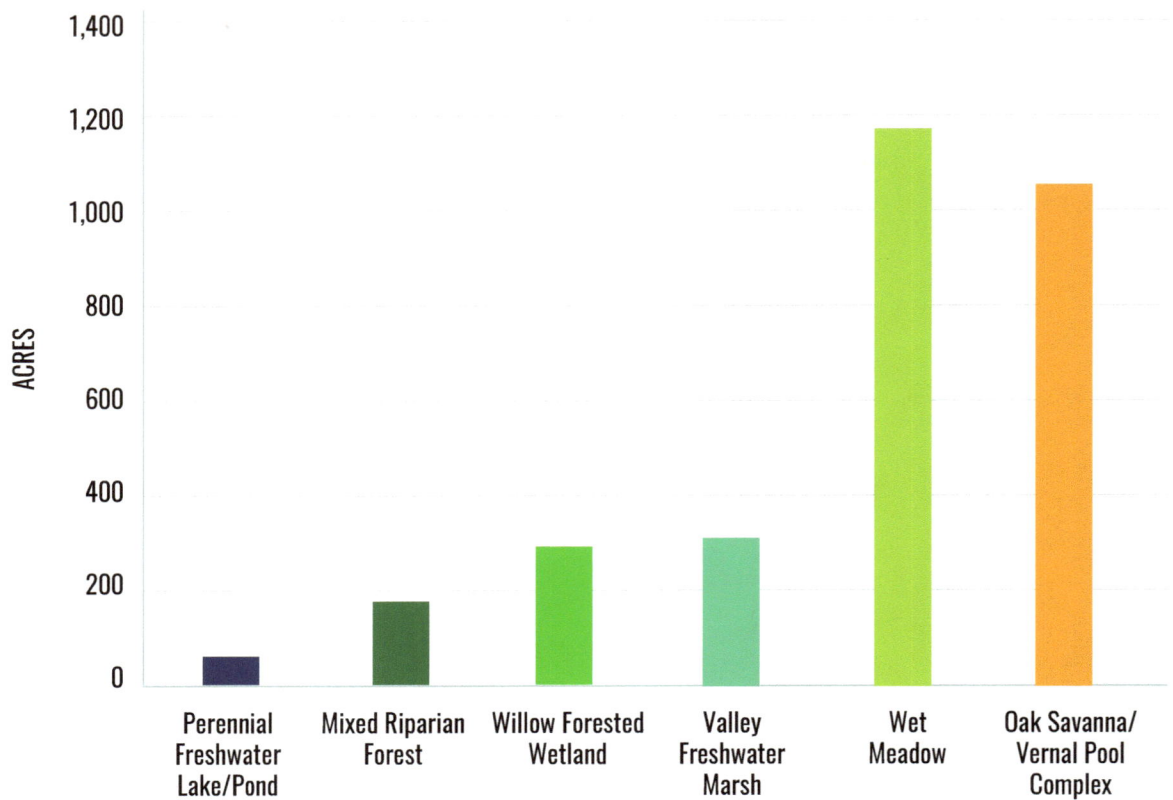

Fig. 5. Total area of each historical (ca. 1850) habitat type mapped within the study area. Perennial freshwater lake/pond, mixed riparian forest, willow forested wetland, and valley freshwater marsh composed the core of the Laguna wetland complex, while extensive areas of wet meadow and oak savanna/vernal pool complex occupied higher surrounding areas.

2008). Lower-elevation areas within the Laguna basin that were permanently flooded supported ponds and perennial wetlands, while slightly higher-elevation areas that were seasonally flooded supported seasonal wetlands. High groundwater levels throughout the lower parts of the Laguna basin supplied a key source of water during the dry season; seeps and springs in the Merced Formation to the west provided an additional source of groundwater input (Cardwell 1958, Nishikawa 2013). Soil configuration also impacted the distribution of particular wetland types: for example, wet meadow often occurred in areas with poorly drained clay soils, while vernal pools occurred in areas with a subsurface hardpan.

The Laguna was a highly productive ecosystem historically, and supported a wide variety of resident and migratory wildlife. While a comprehensive analysis of historical wildlife support is outside the scope of this study, late 19th- and early 20th-century descriptions include glimpses into the many species found in the area's open water, riparian, and wetland habitats, including several species currently listed as threatened or endangered. For example, early observers noted "salmon-trout" (possibly coho salmon or steelhead) in Lake Jonive, along with beaver and California tiger salamander in the Laguna's open water habitats (*Sonoma Democrat* 1879, *The Sebastopol Times* 1903a, Vallejo et al. 2000). An important stop on the Pacific Flyway (Laguna Technical Advisory Committee 1989), the Laguna also supported abundant waterfowl in open water habitats and surrounding tule marshes (Marryat 1855, *Petaluma Journal & Argus* 1869, *Petaluma Weekly Argus* 1881), and provided breeding habitat for Western yellow-billed cuckoo in the brushy "willow bottoms" surrounding the Laguna (Shelton 1911). Grizzly bears and large herbivores such as deer, elk, and pronghorn antelope

BEAVERS IN THE LAGUNA DE SANTA ROSA

Physical controls were not the only factors that shaped the distribution of habitat types within the Laguna historically. Certain plant and animal species, such as beaver, also had a significant influence on landscape patterns. Known as "ecosystem engineers" for their ability to substantially alter ecosystem structure, beavers likely helped to shape the historical Laguna landscape by constructing dams, which in turn altered the flow of water and the movement of sediment and debris through the system. Beaver activity may have played an important role historically in creating and sustaining ponds, marshes, and other wetland habitats within the Laguna, which in turn contributed to overall landscape heterogeneity and species richness (Wright et al. 2002, Lanman et al. 2013).

Estimates of historical beaver density in the region are not available, but early 19th-century sources describe the presence of large numbers of beavers within the Laguna. For instance, in 1833 Mariano Vallejo wrote of "great tulare lakes teeming with beaver [*grandes lagunas tulares, y abunda de castores*]" (Vallejo et al. 2000), while José Figueroa (1834) found the Laguna to have "many beavers." By the mid-1800s, trapping by the Russian American Fur Company and others had largely extirpated beavers from the region (Lanman et al. 2013). Evidence of beaver recolonization within the Laguna watershed has recently been observed in several locations along Santa Rosa Creek, though to date there is no record of beaver re-occupation in the Laguna mainstem (K. Lundquist pers. comm.).

1875 illustration courtesy of Biodiversity Heritage Library, Creative Commons.

frequented the Santa Rosa Plain (Marryat 1855, *Sonoma County Democrat* 1861). While the Laguna continues to provide habitat for many wildlife species, some, such as the Western yellow-billed cuckoo, have been extirpated over the past century and a half (Honton and Sears 2006).

HYDROLOGY

Channel Network

The Laguna mainstem channel was relatively shallow and narrow throughout much of the study area historically. In many places, the Laguna flowed through a shallow network of sloughs that threaded through the wetland complex. Nineteenth and early twentieth century surveyors and observers noted the poorly defined nature of the Laguna channel in a number of locations throughout the study area. For example, just south of present-day Highway 12, GLO surveyor Thomas Whitacre described the Laguna mainstem as a "swale" just over 30 feet wide, suggesting that it was quite shallow (Whitacre 1853; Fig. 6). Sarepta Ann Turner Ross, an early pioneer whose family came west in a wagon train and moved to Sebastopol in 1854, recalled that near the present-day Occidental Road crossing "there were no channels, and the water spread out for quite a long distance" (Ross 1914). The lack of a channel in this location necessitated the construction of a "squatter's bridge" made by "[hauling in] logs until they could cross over without getting stuck in the mud" (Fig. 7). In some areas the mainstem channel may have been more well-defined, though still relatively shallow: just north of the confluence with Irwin Creek, for instance, GLO surveyor Seth Millington described the mainstem as "a channel about 3 feet deep" (Millington 1865).

Fig. 6. (right) 1907 Petaluma & Santa Rosa Railway map showing narrow "main channel" of Laguna under HWY 12, set within the broader wetland complex. Calder Creek is shown flowing into the Laguna from the west. *(P&SRR Co. 1907, courtesy of Sonoma County Library)*

Fig. 7. (left) "Squatters bridge," 1859. The absence of a defined channel near the present-day Occidental Road crossing necessitated the construction of an informal crossing made of logs to avoid getting stuck in the broad, muddy wetland complex. *(Tracy 1859b, courtesy of The Bancroft Library, UC Berkeley)*

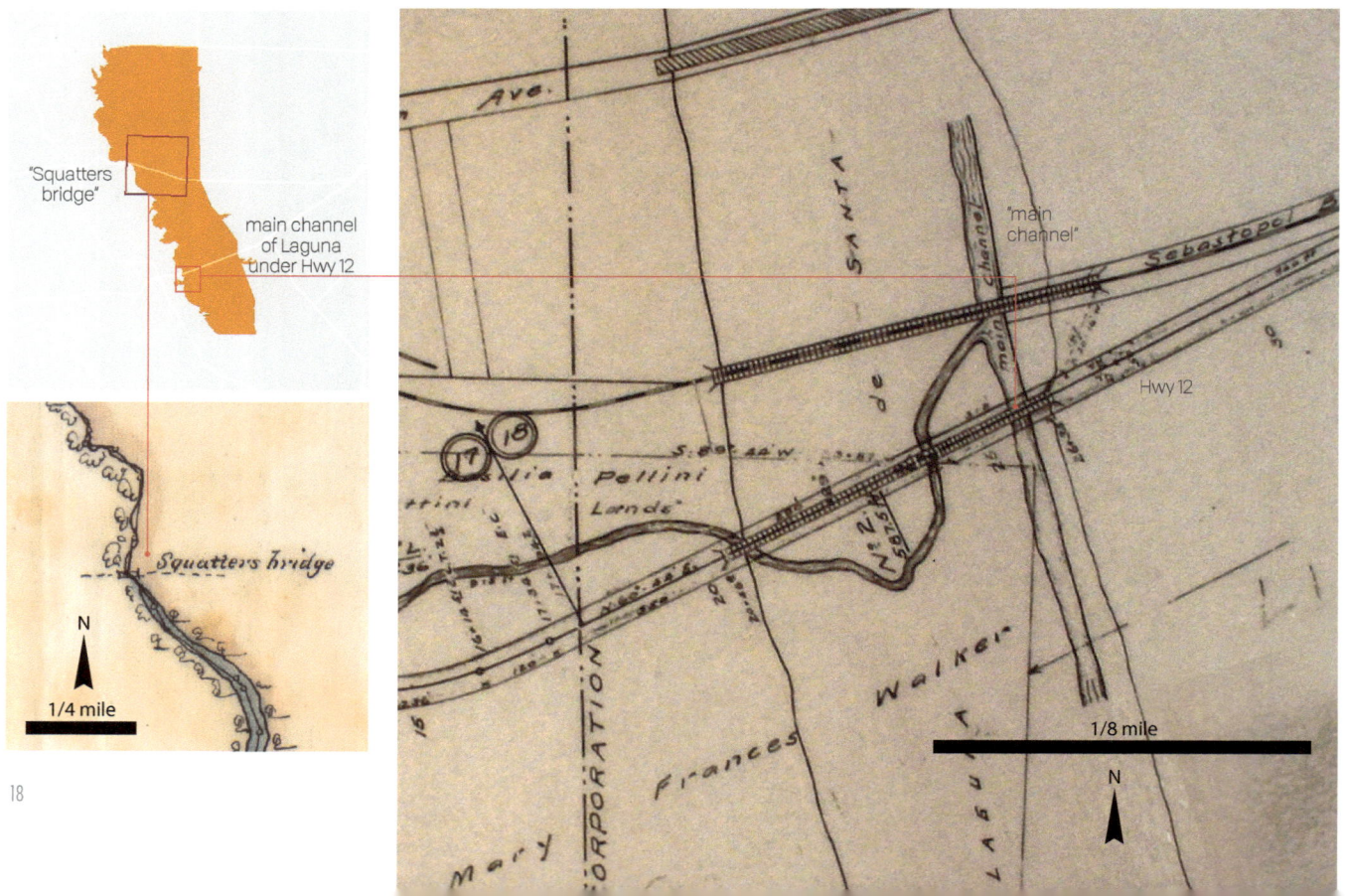

Numerous tributaries flowed into the Laguna wetland complex from both the Santa Rosa Plain to the east and the hills of the Wilson Grove Formation to the west. All of the larger tributaries, including Santa Rosa, Irwin, Duer, Gravenstein, and Roseland creeks, drained off of the Santa Rosa Plain. The floodplains surrounding the lower reaches of these creeks supported seasonal wetlands such as wet meadow, as well as limited areas of perennial freshwater marsh and mixed riparian forest.

Santa Rosa Creek, which originates in the Mayacamas Mountains east of Santa Rosa, is the largest tributary to the Laguna. During the mid-19th century the mainstem connected with the Laguna just south of present-day Guerneville Road, though the Santa Rosa Creek channel would likely have shifted considerably across its alluvial fan on a decadal or multidecadal timescale. Many 19th-century maps (e.g., Tracy 1859b, U.S. Surveyor General's Office 1866, Bowers 1867, Thompson 1877b) depict Santa Rosa Creek with a single channel within the study area, while early to mid-20th-century sources (e.g., USGS [1933-5]1942, USDA 1942) show several distinct branches or distributaries approaching the Laguna (Fig. 8). It is unknown whether the multiple channels shown in these sources are representative of earlier conditions, or whether they reflect changes in channel morphology resulting from direct modifications (e.g., ditching) or changes in upstream sediment and water supply. Further research into early land-use changes with the Santa Rosa Creek subwatershed may shed light on alterations in sediment supply and potential impacts on channel morphology during the 19th century.

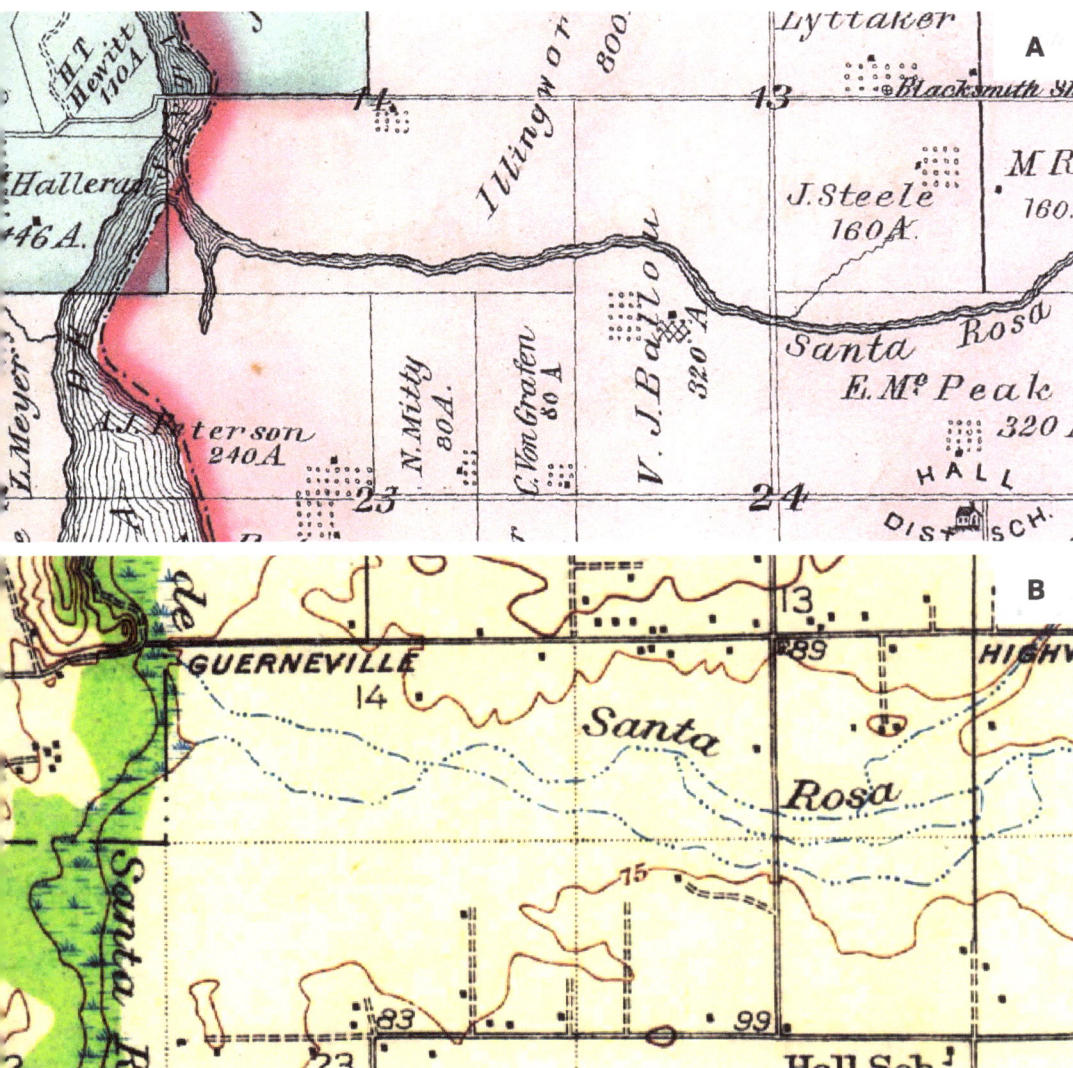

Fig. 8. While many 19th-century maps depict Santa Rosa Creek's confluence with the Laguna as a single channel (e.g., A), early- to mid-20th-century sources tend to show multiple channels approaching the Laguna (B). It is unknown whether these differing representations are reflective of changes in channel morphology or differences in cartography. (A: Thompson 1877b, courtesy of David Rumsey Map Collection; B: USGS [1933-5]1942)

N

1/2 mile

Flooding and Flow Variability

Historically, the extent of inundation in the Laguna wetland complex varied dramatically throughout the year. During wet-season floods, shallow open water likely covered much of the mapped wetland area (Fig. 9). Accounts of late 19th-century floods, for instance, describe flooding between half a mile and several miles wide (e.g., *Daily Alta California* 1866, *Daily Courier and Petaluma Imprint* 1895 in Cummings 2006).

In addition to flows originating within the watershed upstream, the Laguna was also affected by flooding on the Russian River downstream. As a result of the Laguna's gradual slope, floodwaters (and fine sediment) from the Russian River often backed up into the Laguna watershed during high flows, increasing the extent of flooding within the Laguna wetland complex. The substantial flood storage capacity within the Laguna has been shown to significantly reduce flows and flood heights on the Russian River downstream of the Laguna-Mark West Creek confluence (SCFCWCD 1965, Sloop et al. 2007, Curtis et al. 2013).

By late summer, the extent of open water would once again be mostly confined to Lake Jonive, though small pools and shallow surface water likely persisted in other parts of the wetland complex. The lack of perennial surface flow, and the existence of isolated pools of standing water during the dry season, is reflected in a number of early descriptions of the Laguna. For instance, while surveying just north of Lake Jonive in February 1865, GLO surveyor Seth Millington noted, "The greater part that is covered with water at this time is dry at other seasons of the year" (Millington 1865). Writing of his travels across the Santa Rosa Plain to Bodega in September of 1810, Gabriel Moraga described "a lagoon and a stream with many pools of retained water [*una laguna y un arroyo con muchas posas de agua retenida*]" (apparently the Laguna mainstem); he contrasted this with "another stream with some running water [*otro arroyo con alguna agua corriente*]" (apparently Santa Rosa Creek), suggesting that standing water was abundant in ponds but that surface flow in the Laguna was minimal or nonexistent by that point in the dry season, at least in that year (Moraga 1810, Priestly 1946).

"During severe Winters this Laguna spreads out and overflows a large portion of the surrounding country."
— Menefee 1873

Fig. 9. The Laguna complex was frequently inundated by wet-season flood events, which often covered much of the study area with shallow open water. This series of images depicts a series of mid-20th-century floods. **Opposite:** Flooded pastures near Sebastopol in 1951. **Top:** Floodwaters at Occidental and High School roads during the 1951 flood. **Center:** Flooding on Highway 12 near the Laguna, ca. 1940. **Bottom:** Flooding on Santa Rosa Creek just upstream of the Laguna confluence, December 19, 1955. *(Opposite: Photo #13266, courtesy of Western Sonoma County Historical Society; Top: Photo #13219, courtesy of Western Sonoma County Historical Society; Center: Photo #2036, courtesy of Western Sonoma County Historical Society; Bottom: C-1967-6, courtesy of Sonoma County Water Agency)*

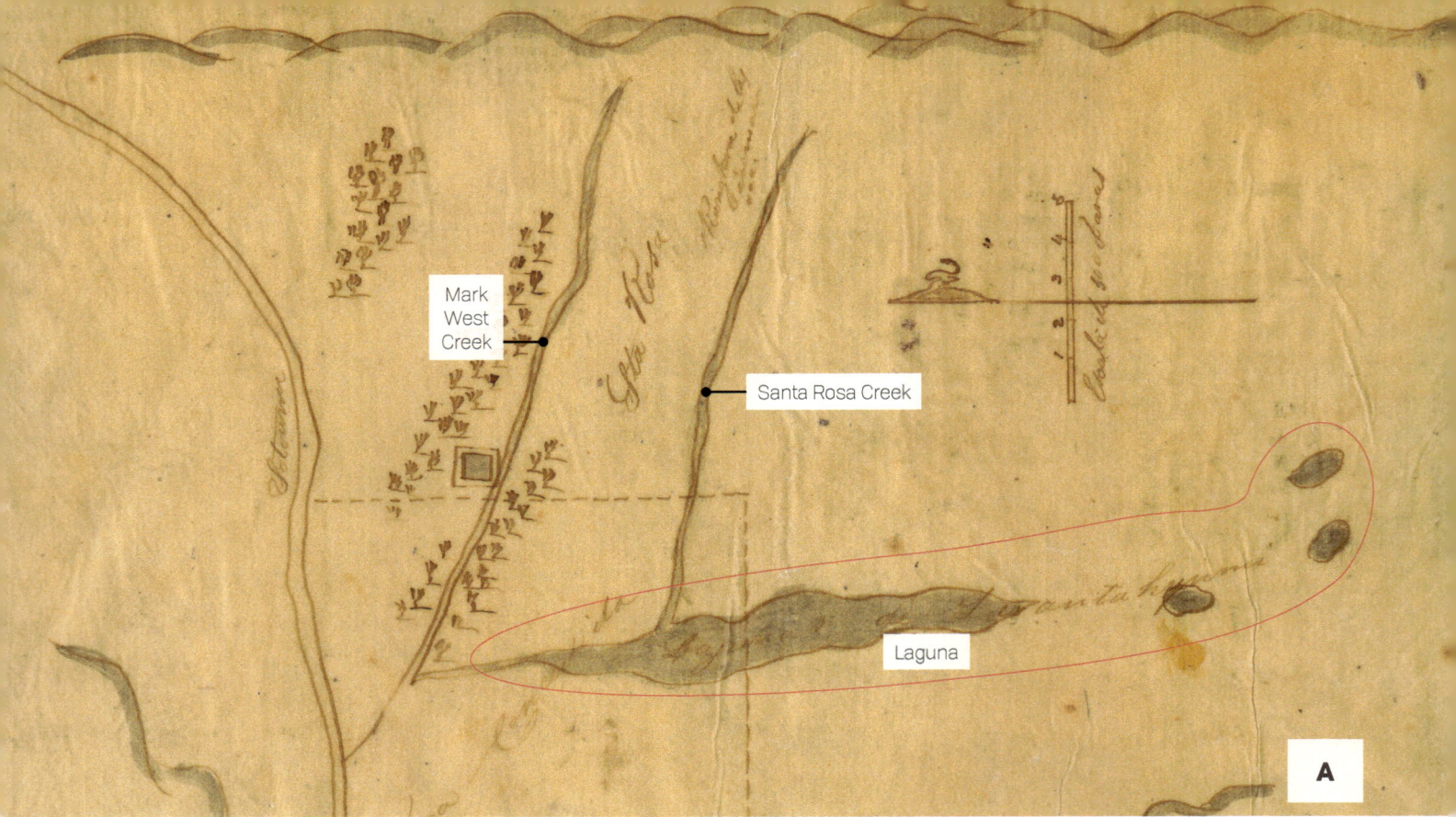

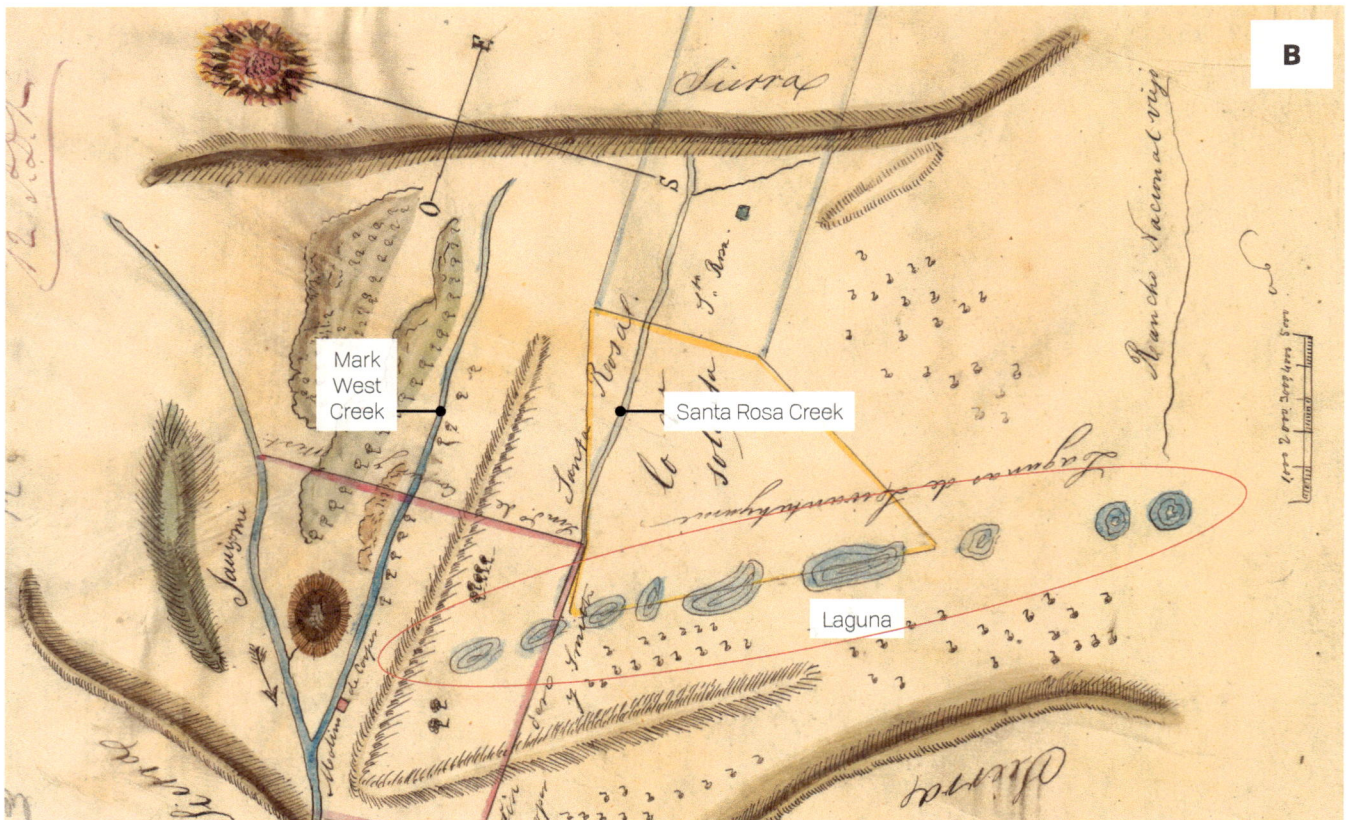

Fig. 10. **Early land grant maps,** such as these maps from ca. 1840, show the Laguna as a series of perennial ponds or lakes. The lower map labels it "Laguna de Livantuligüeni"; Livantuligüeni (there are numerous spelling variants, including Livantolomi and Livancacayomi) refers to an Indian rancheria located just west of the Laguna (Merriam 1966, Vallejo et al. 2000). *(A: USDC ca. 1840a; B: USDC ca. 1840b; both courtesy of The Bancroft Library, UC Berkeley)*

Moraga's description is echoed in later accounts that describe the Laguna in this area as "a series of lakes" (Thompson 1877a) and "a series of ponds and wide marshes" (Watson et al. 1917). Several early land grant maps likewise depict the Laguna as a string of ponds (Fig. 10).

Like the Laguna mainstem, streamflow in Santa Rosa Creek and other tributaries was also highly seasonal. In Santa Rosa Creek, surface flow during the dry season typically did not extend to its confluence with the Laguna (Fig. 11): an 1889 report stated that "during the summer months both [Santa Rosa and Mark West creeks] are lost in the Santa Rosa plain" (The Lewis Publishing Company 1889).

Springs and high groundwater levels were likely a primary source of water for many parts of the Laguna wetland complex during the dry season, maintaining year-round saturated soil in lower elevation areas and providing a source of water for perennial ponds. The earliest comprehensive survey of groundwater conditions in the Santa Rosa Plain and surrounding areas, completed in 1951 (which the report describes as being prior to substantial groundwater overdraft), found that during the dry season "the watertable [sic] is at about the same level surface as that of the Laguna," and that groundwater likely sustained high evapotranspiration rates by "reeds, tules, willows, and other water-loving plants which flourish around the margins of the Laguna" (Cardwell 1958). In addition, the study found that "much ground water is discharged from the Merced formation on the western side of Santa Rosa Valley through springs and seeps."

Fig. 11. Surface flows in Santa Rosa Creek, the largest tributary to the Laguna, did not extend to the Laguna confluence during the dry season. This 1879 photograph, looking towards Main Street Bridge in Santa Rosa (upstream of the study area), shows the dry bed of Santa Rosa Creek. (Annex Photo 546, courtesy of Sonoma County Library)

Valley Freshwater Marsh

Valley freshwater marsh occupied approximately 130 ha (310 ac) within the study area. The largest concentration of marsh occurred in the northern part of the study area, though several smaller patches existed near the confluences with Irwin Creek and Duer Creek and along the Laguna mainstem channel in the southern part of the study area. The marsh was semi-permanently to perennially flooded, and likely dominated by plants such as tule, bulrush, and cattail.

Early 19th- through early 20th-century accounts suggest that perennial, tule-dominated marshes were a prominent habitat type within the Laguna basin. As noted above, General Vallejo observed "great *tulare* lakes teeming with beaver" in 1833 (Vallejo et al. 2000). In the early 20th century, Holway (1913) noted that "a series of lagoons and tule marshes mark the channel of the sluggish current" between Sebastopol and Trenton. Early soil surveys also support the presence of extensive areas of emergent marsh. The earliest soil map of the area, completed in 1915, depicts large areas using the tufted "marsh" symbol (Fig. 12). This symbol is a fairly generalized indicator of wetland presence, and does not in itself represent direct evidence for the presence of valley freshwater marsh (it may also indicate other types of "swamp land"; USGS 1913). However, in conjunction with other

Fig. 12. Soil map of the study area (outlined in black), 1915. The overall extent of freshwater marsh and perennial forest is depicted here using the tufted "marsh" symbol, primarily in areas of Yolo silty clay loam (Yc). *(Watson et al. 1915, courtesy of University of Alabama)*

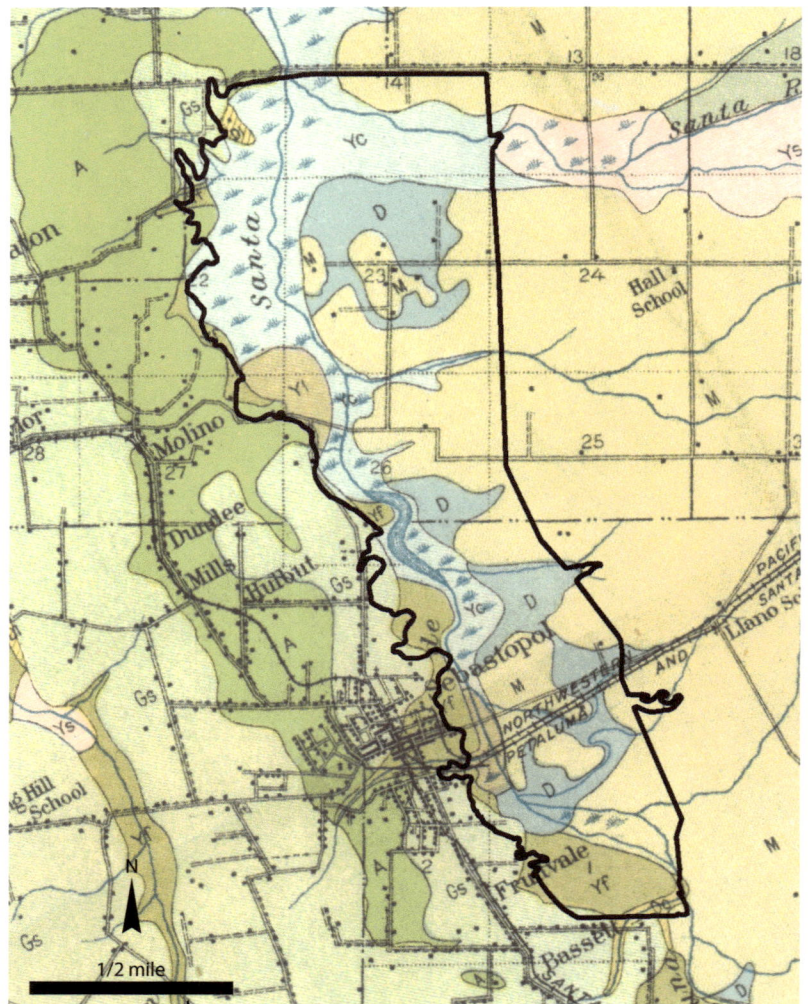

evidence the presence of the marsh symbol lends support to the Valley Freshwater Marsh classification in many areas. The accompanying soil report states, "Near Sebastopol [the Laguna de Santa Rosa] flows through a series of ponds and wide marshes" (Watson et al. 1917).

The largest area of freshwater marsh within the study area occurred north of Occidental Road, where it occupied approximately 100 ha (260 ac), mostly on the eastern side of the Laguna mainstem (Fig. 13). Early surveyors documented the presence of tules throughout this region (Tracy 1859a, Millington 1865), as well as shallow standing water (Millington 1865; Fig. 14).

Fig. 13. (left) This 1859 survey plat of Rancho Llano de Santa Rosa shows valley freshwater marsh, depicted with a marsh symbol and labeled "tules," on the east side of the channel in the northern part of the study area. On the west side of the channel, willow forested wetland (page 26), is depicted with a tree symbol and labeled as a "swamp". (Tracy 1859b, courtesy of The Bancroft Library, UC Berkeley)

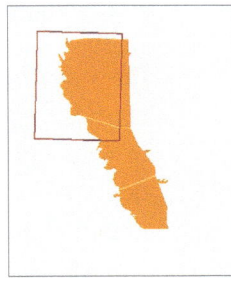

Fig. 14. (right) Summary of General Land Office survey evidence for valley freshwater marsh. GLO surveys conducted in August 1859 and February 1865 noted tules, swamp, and shallow open water when traversing the Laguna.

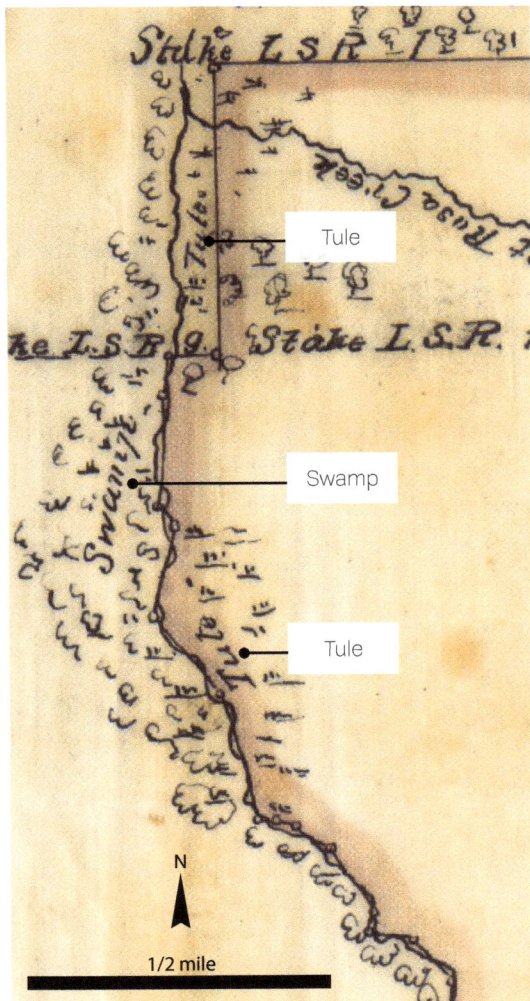

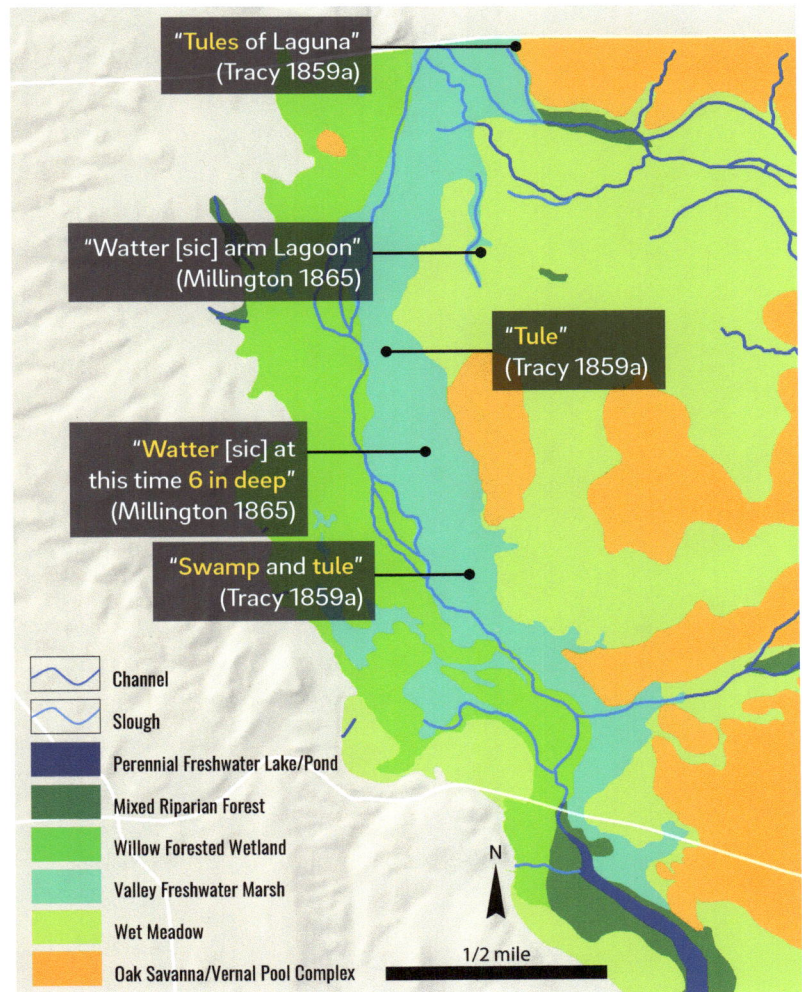

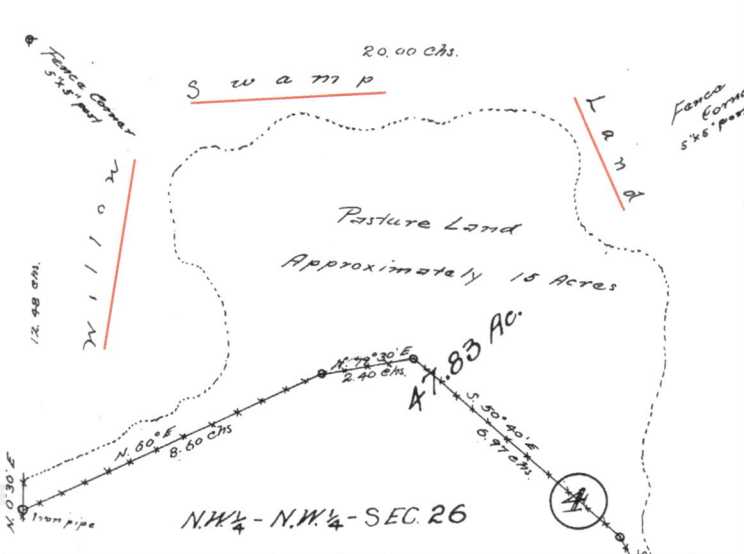

Willow Forested Wetland

The northwestern side of the study area supported an extensive willow forested wetland which occupied approximately 120 ha (300 ac). This wetland type was dominated by willows, with tules, Oregon ash, and other species also present. Though less permanently flooded than the valley freshwater marsh to the east, the low gradient and lack of defined channels through this area resulted in frequent (seasonal to semi-permanent) shallow flooding. On its southern end, the willow forested wetland graded into mixed riparian forest; the boundary between these two habitat types was difficult to determine precisely, and it was delineated primarily based on proximity to Lake Jonive (see page 27).

The willow forested wetland is depicted in early sources as a forested riparian wetland area, and was often described by early surveyors as a "swamp" (Fig. 15; see also Fig. 13 on page 25; Tracy 1859b, Martin and Eliason 1861b, Millington 1865, Mueller 1921). While historical sources support the general east-west separation in the distribution of willow forested wetland and valley freshwater marsh (e.g., Tracy 1859b, USDA 1942), in reality there was likely a substantial degree of heterogeneity between these two wetland types resulting from local differences in topography and hydrology. For instance, while traveling west across the Laguna just north of the Irwin Creek confluence in February 1865, GLO surveyor Seth Millington noted, "The line through the Lagoon is mostly covered with a growth of willow and ash timber interspersed with tule," suggesting considerable intermixing between habitat types (Millington 1865).

Willow forested wetland is visible in the 1942 aerial photos as a densely forested expanse (Fig. 16). Although early to mid-20th-century habitat patterns were not explicitly analyzed in this study, it appears that by 1942 this wetland type may have become substantially more extensive than it was during the mid-19th century, perhaps as a result of ditching and draining that had occurred by this time, reducing the frequency of inundation. Significant areas mapped as valley freshwater marsh appear to have converted to willow forested wetland by the mid-20th century.

Fig. 15. (top) This 1921 subdivision map, which covers only the southern part of the willow forested wetland, labels it as "willow swamp land." *(Mueller 1921, courtesy of Sonoma County Recorder)*

Fig. 16. (bottom) Willow forested wetland appears to have expanded considerably by the time the 1942 aerial photos were taken. *(USDA 1942)*

Perennial Freshwater Lake/Pond

One of the central features within the study area was a deep, perennial freshwater lake known as Lake Jonive. Located east of present-day Sebastopol, this elongate lake extended from just north of present-day Highway 12 to present-day Occidental Road, approximately 3 km (2 mi) north. The lake was a popular boating and swimming destination during the late 19th and early 20th centuries, as documented in numerous landscape photographs from the period (Fig. 17).

Lake Jonive was fairly narrow, ranging from approximately 40–80 m (150–250 ft) wide (Tracy 1859b, Dyer 1861, Millington 1865, *The Sebastopol Times 1903b*). However, compared with other parts of the Laguna wetland complex, which were typically quite shallow, Lake Jonive was relatively deep in areas. After taking soundings of the lake in the early 1900s, a group of high school students from Sebastopol reported:

> "Beginning at the south end, in a narrow slough, we found the first depth to be 5 ½ feet. At the place where the slough broadened into the larger part the depth was 9 feet. In 100 yards' distance the depth increased gradually to 23 feet... The greatest depth was 23 ½ feet." –Holway 1913

Fig. 17. Boating on Lake Jonive, ca. 1907. The lake was a popular swimming and boating destination during the late 19th and early 20th centuries. *(CHS-45682, courtesy of USC Libraries, California Historical Society Collection)*

Lake Jonive was bordered by a mixed riparian forest dominated by oaks and willows (see page 30). Landscape photographs from the late 19th and early 20th centuries also appear to show a narrow band of tules around the margins of the lake (not mapped separately; Fig. 18).

While conclusive evidence for other perennial ponds within the study area was not found, it is likely that a number of other small, shallow open water features existed in depressions within perennial wetlands such as valley freshwater marsh. Several other perennial freshwater lakes existed within the Laguna wetland complex outside of the study area, most notably Ballard Lake, located north of Guerneville Road (Watson et al. 1915, Heintz 1990, Denner 2002, Cummings 2004). In the early 20th century, Holway (1913) reported that Ballard Lake was approximately 0.4 km (0.25 mi) long, 75 m (250 ft) wide, and up to 7.5 m (25 ft) deep.

Fig. 18. Though relatively narrow, Lake Jonive extended along the valley for approximately 3 km (2 mi) and reached a maximum depth of just over 7 m (23 ft). As shown in these early 20th-century photographs, the lake was bordered by tules and a mixed riparian forest dominated by oaks and willows. (Above: Photo #2027, courtesy of Western Sonoma County Historical Society; Opposite, top: Photo #1376, courtesy of Laguna de Santa Rosa Foundation; Opposite, bottom: Photo #2028, courtesy of Western Sonoma County Historical Society)

"From the clear waters of [Lake Jonive] have been caught salmon-trout that filled the sportsman's heart with joy."

–The Sebastopol Times 1903a

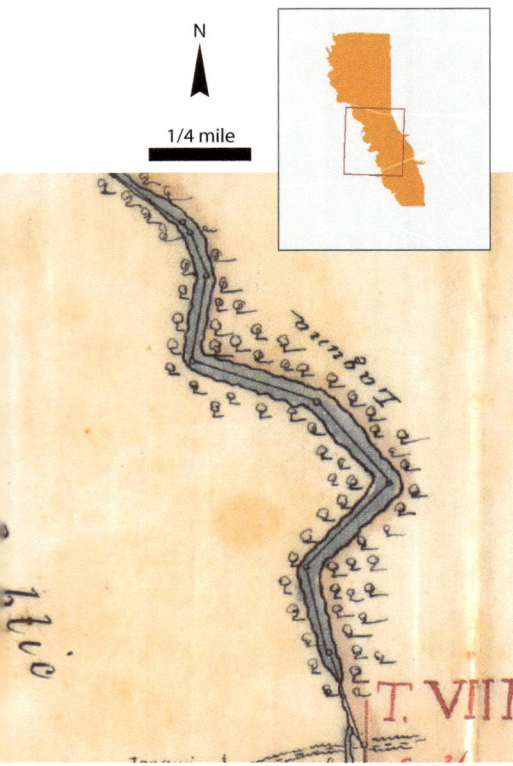

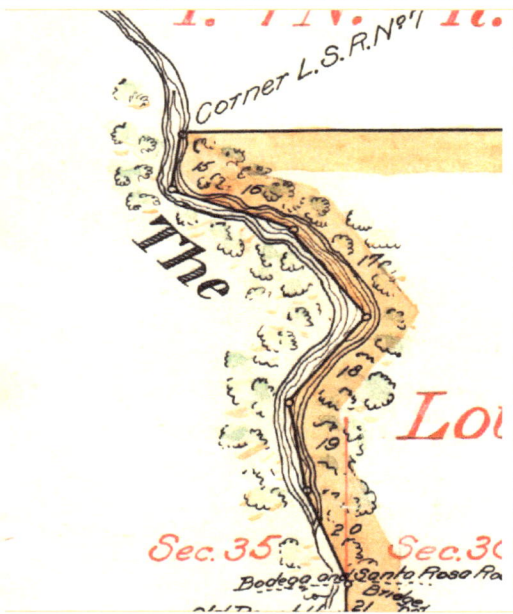

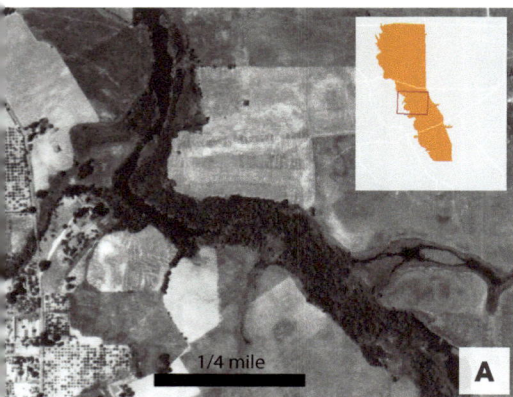

Mixed Riparian Forest

Mixed riparian forest, dominated by oaks and willows, occupied 70 ha (180 ac) within the study area. This habitat type occurred primarily along the margins of Lake Jonive, where it covered approximately 60 ha (150 ac) and varied in width from less than 10 m to over 350 m (30–1,150 ft; Fig. 19). Turn-of-the-century newspaper articles described the lake as "bordered by a heavy growth of oak" and as "bordered with oaks, willows, etc" (*The Sebastopol Times* 1903a, 1903b; see also Fig. 18 on pages 28-29). Oregon ash and boxelder were also likely present within these riparian areas (Figueroa 1834, Millington 1865, Shelton 1911, Howell 1951, Waaland 1989). By the time the earliest aerial photos were taken in 1942, portions of the forest had already been cleared, but remnants are still visible (USDA 1942; Fig. 20A).

At its upstream end, just south of Lake Jonive, the riparian forest assumed a lower growth form dominated by smaller trees and shrubs: GLO surveyor Thomas Whitacre described the forest in this area as a "dense brushy thicket" (Whitacre 1853; Fig. 21 and Fig. 22). Small patches of riparian forest also occurred near the Laguna mainstem channel in the southern part of the study area.

Conclusive historical evidence for extensive riparian forests along tributaries within the study area was not found, though tributaries like Santa Rosa Creek and Irwin Creek appear to have supported small patches or narrow corridors of mixed riparian forest (Fig. 20B,C). For example, the 1942 aerial photos show narrow bands of trees bordering the Santa Rosa Creek channels, though much of the surrounding area was under cultivation by that time and it is possible that substantial numbers of riparian trees had already been removed. Most parts of the riparian corridor along Santa Rosa Creek visible in the 1942 aerial photos are between 10–30 m (30–100 ft) in width, though a corridor up to 60 m (200 ft) wide is visible just upstream of the confluence with the Laguna, perhaps representing a remnant of a more extensive

Fig. 19. (left, top and bottom) Riparian forest along Lake Jonive, ca. 1860. These early maps depict riparian forest along the margins of Lake Jonive. *(Top: Tracy 1859b, courtesy of The Bancroft Library, UC Berkeley; Bottom: Dyer 1861, courtesy of Bureau of Land Management)*

Fig. 20. (below) Details from 1942 aerial photomosaic showing patches of riparian forest along Lake Jonive (A) and tributaries (B,C). Lake Jonive supported a broad (10–350 m wide) riparian forest, while tributaries such as Santa Rosa Creek and Irwin Creek appear to have supported small patches or narrow corridors of forest. *(USDA 1942)*

riparian forest. On the other hand, some early sources suggest an absence of broad riparian forests along this portion of Santa Rosa Creek. For example, while surveying within the "bottom land" of Santa Rosa Creek along the eastern edge of the study area in February 1865, Millington reported, "No

Fig. 21. (top) This ca. 1905 image of the Petaluma and Santa Rosa Railway trestle over the Laguna shows the riparian forest south of Lake Jonive, where it was dominated by smaller trees and shrubs. *(Photo #6861, courtesy of Western Sonoma County Historical Society)*

Fig. 22. (bottom) Panorama of Sebastopol looking eastward across the Santa Rosa Plain, 1905. Riparian forest marking the course of the Laguna is visible in the background at the top right of the image. *(Photo #3462, courtesy of Western Sonoma County Historical Society)*

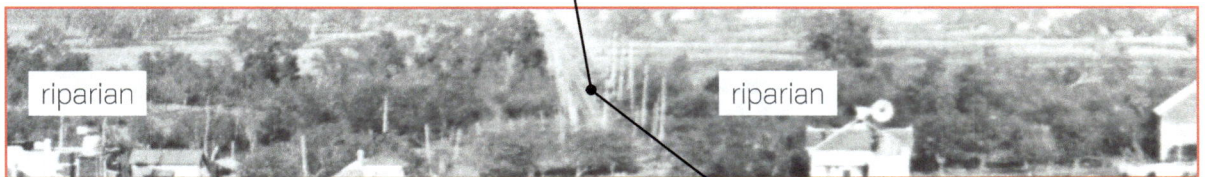

tree near for bearings." A short distance upstream he again noted, "Very little timber. Ash and white oak" (Millington 1865). It appears likely that much of the floodplain along Santa Rosa Creek within the study area was dominated by seasonal wet meadow rather than riparian forest (see page 32).

Wet Meadow

Wet meadow was prevalent throughout the study area historically, covering 470 ha (1,170 ac). This habitat type occupied temporarily or seasonally flooded floodplain areas adjacent to the perennial wetlands and riparian habitats in the center of the Laguna wetland complex. Wet meadows were characterized by poorly drained, clay-rich soils and an herbaceous plant community likely composed of a mix of grasses, rushes, and forbs.

Early maps and textual accounts often do not document or describe seasonal wetlands like wet meadow, likely as a result of their variable and somewhat ephemeral nature. As a result, the evidence for wet meadows within the Laguna wetland complex is much more limited than for other wetland types. Many of the areas mapped as wet meadow correspond approximately to areas mapped as Dublin Clay Adobe by the 1915 soil survey (Watson et al. 1915; see Fig. 12 on page 24). Dublin Clay Adobe is described as having "incomplete" drainage (Watson et al. 1917). There is also evidence for wet meadows from the 1942 aerials in the form of darker patches in areas with herbaceous cover, which are often a signature of wetter conditions and poor drainage (Fig. 23).

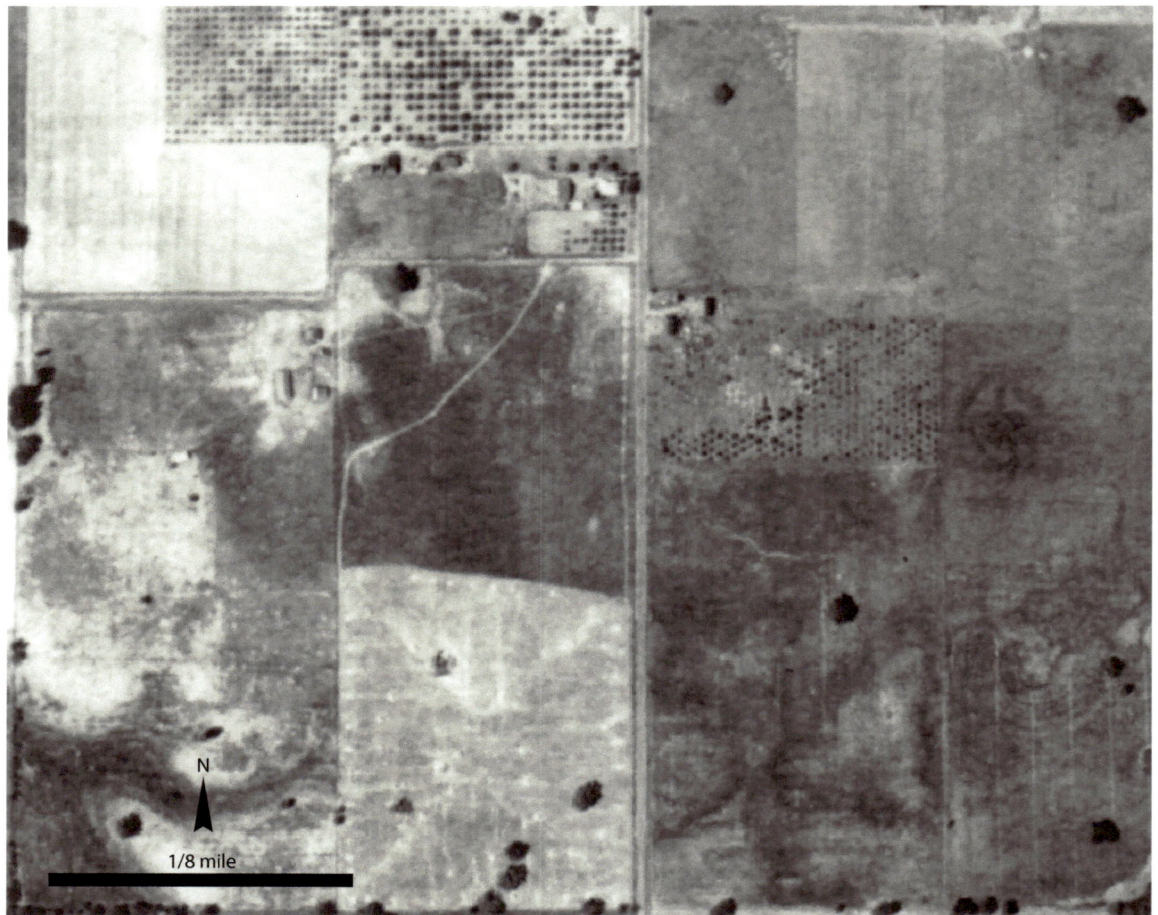

Fig. 23. Wet meadow, 1942. This detail from the historical aerial photomosaic east of the Laguna (just north of Irwin Creek) shows the dark patches often characteristic of wetter conditions and poor drainage. *(USDA 1942)*

Oak Savanna/Vernal Pool Complex

Slightly higher-elevation areas surrounding the wet meadow supported oak savanna/vernal pool complex (Fig. 24). This habitat type was dominated by a sparse cover of oak trees (mostly valley oak) intermixed with an array of vernal pools within a grassland matrix. Previous analyses (e.g., Waaland 1989, Waaland 1990, Butkus 2011b), along with evidence from historical soil surveys, aerial photos (Fig. 25), and other archival sources, suggest that oak savanna/vernal pool complex historically occupied large areas of the Santa Rosa Plain east of (and contiguous with) the study area. The mapped area thus represents a small fraction of a much more extensive oak savanna/vernal pool complex within the region historically.

Oak savanna/vernal pool complex largely occurred on areas shown as Madera Loam in the 1915 soil map (Watson et al. 1915; see Fig. 12 on page 24), which was characterized by mounded topography and an impermeable subsoil layer—features often associated with vernal pool formation. The accompanying soil report (Watson et al. 1917) describes Madera Loam as follows:

> "The surface is usually uneven, as the result of the occurrence of numerous small mounds and intervening depressions. These retain water during the rainy season, owing to the impervious subsoil and hardpan... The original growth on this soil consisted of grasses and scattered valley oaks."

Numerous surveyors recorded the presence of "white oaks" (valley oaks), black oaks, and "scattering oak timber" on the eastern side of the study area (Fig. 26; Whitacre 1853, Gray 1857, Millington 1865). In contrast, vernal pool complexes, like wet meadow and other seasonal wetland types, were generally not mentioned in early surveys.

Fig. 24. Scattered oaks and vernal pools near Gravenstein Creek, 1942. *(USDA 1942)*

Fig. 25. Vernal pool/ swale complexes on the Santa Rosa Plain. Scattered oak trees are also visible. (top) Near Arlington Avenue and Scenic Avenue, February 26, 1966. (bottom) Near Ludwig Avenue and Stony Point Road, February 26, 1966. *(Top: Negative #321-23; Bottom: Negative #321-20; both courtesy of Sonoma County Water Agency)*

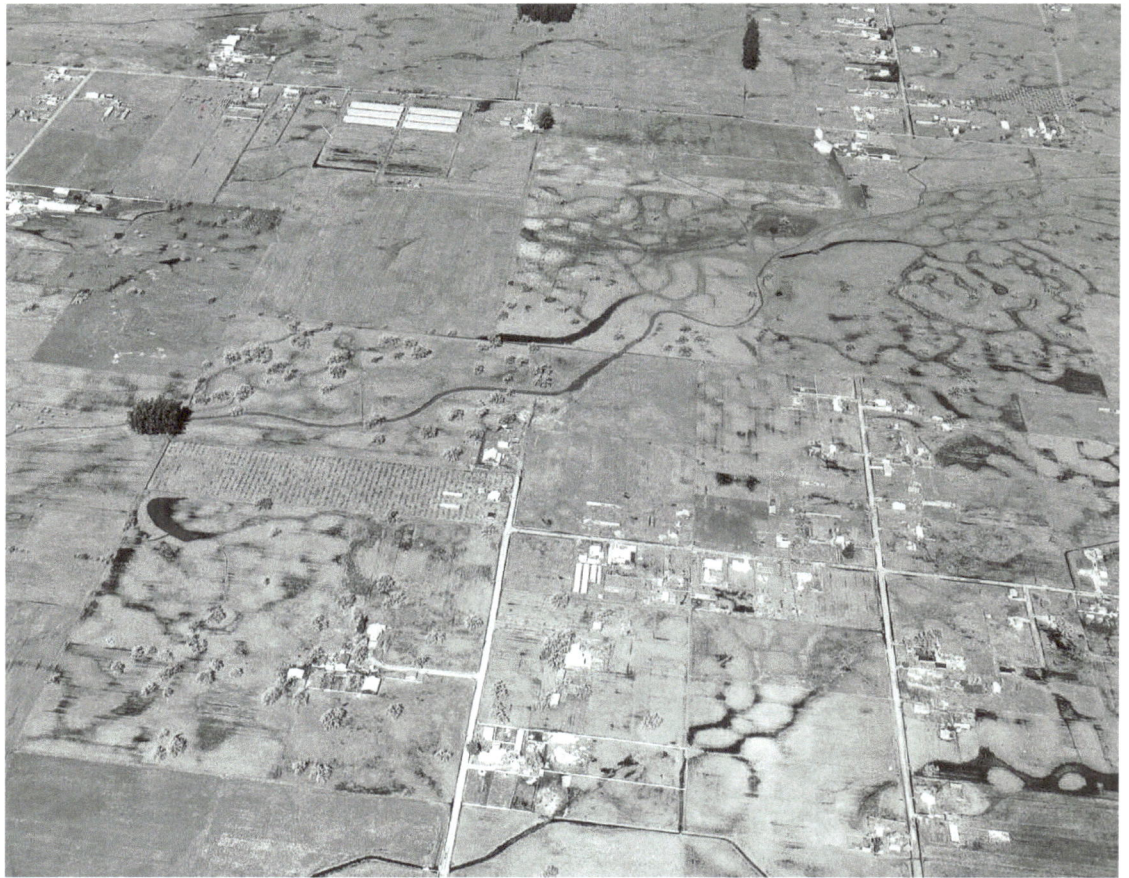

"White oak"
(Gray 1857)

"White oak[s]"
(Millington 1865)

"White oak[s]"
(Millington 1865)

"White and black oak"
(Millington 1865)

"Scattering oak timber"
(Whitacre 1853)

"Dead oak"
(Martin and Eliason 1861a)

"The grand old oaks, with their dark green foliage, stand scatteringly all through the town [Santa Rosa], over the plains, on the slopes and the hill tops, and impart a pleasing and peaceful aspect to the scenery."

– Daily Alta California 1872

Channel
Slough
Perennial Freshwater Lake/Pond
Mixed Riparian Forest
Willow Forested Wetland
Valley Freshwater Marsh
Wet Meadow
Oak Savanna/Vernal Pool Complex

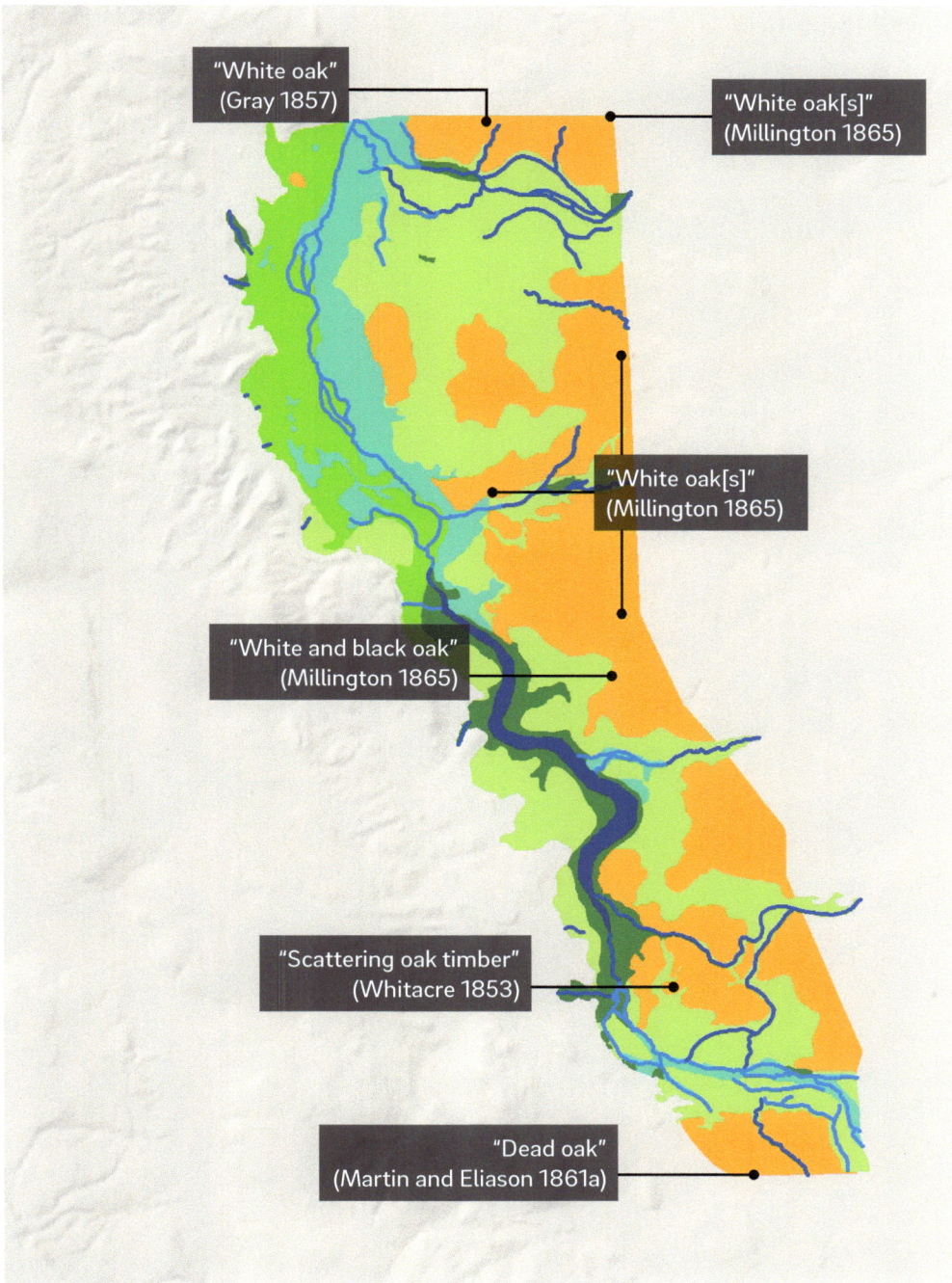

Fig. 26. Summary of county and General Land Office survey evidence for oak lands in the study area. The surveys, conducted between 1853 and 1865, describe valley oaks and black oaks on the plain east of the Laguna.

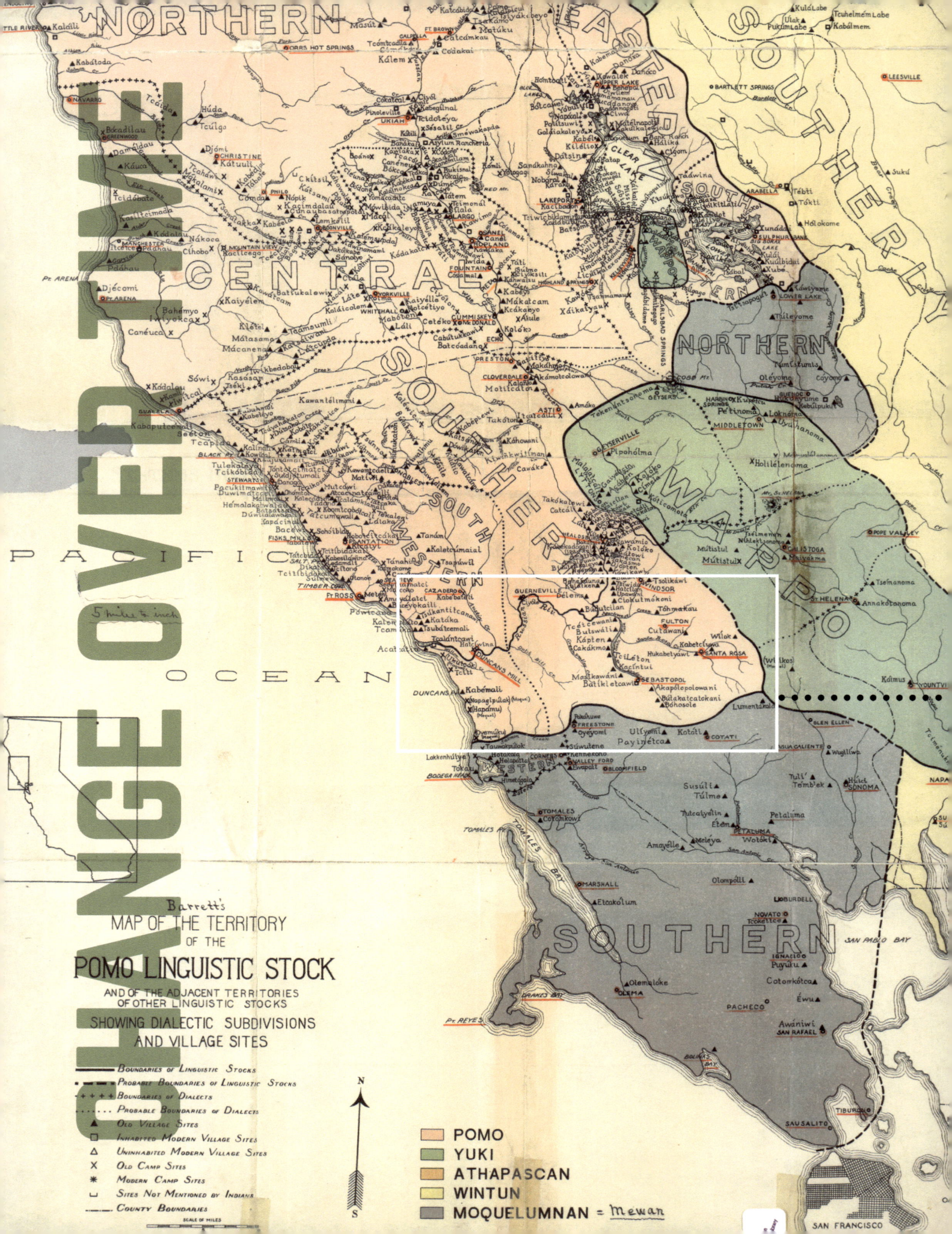

CHANGE OVER TIME

Region / stock labels (as printed on the map):

NORTHERN · CENTRAL · EASTERN · SOUTHERN · SOUTHEASTERLY · WARPO · SOUTH WESTERN · WESTERN · SOUTHERN

PACIFIC OCEAN

Barrett's
MAP OF THE TERRITORY
OF THE
POMO LINGUISTIC STOCK
AND OF THE ADJACENT TERRITORIES
OF OTHER LINGUISTIC STOCKS

SHOWING DIALECTIC SUBDIVISIONS
AND VILLAGE SITES

— — — Boundaries of Linguistic Stocks
— · — · Probable Boundaries of Linguistic Stocks
+ + + + Boundaries of Dialects
· · · · · Probable Boundaries of Dialects
▲ Old Village Sites
□ Inhabited Modern Village Sites
△ Uninhabited Modern Village Sites
× Old Camp Sites
✳ Modern Camp Sites
⊔ Sites Not Mentioned by Indians
— Ⅼ County Boundaries

SCALE OF MILES

N / S

5 miles 5 inch

POMO
YUKI
ATHAPASCAN
WINTUN
MOQUELUMNAN = *Mewan*

San Francisco · San Pablo Bay · Tomales Bay · Bodega Bay · Drakes Bay · Pt. Reyes · Pt. Arena · Little River

OVERVIEW OF LAND USE CHANGES

The Laguna region is believed to have been inhabited by humans for at least the past 7,000 years (Origer and Frederickson 1980), though there is unpublished evidence indicating possible human presence far earlier (T. Origer pers. comm.). At the time of earliest European contact, a number of Pomo villages were located near the Laguna, which provided a productive resource for food and materials for these communities (Fig. 27; Barrett 1908b, Kroeber 1925, Origer and Frederickson 1980, Fredrickson and Markwyn 1990). The Pomo actively managed the landscape in a number of ways, including using fire to alter vegetation cover and harvesting tules for canoes, baskets, and huts (Fig. 28; Anderson 2005, T. Origer pers. comm.). While the precise scale and effects of Native land management practices are open to debate, these activities undoubtedly helped to shape the landscape encountered by Euro-American settlers in the early 1800s.

Notable landscape and ecosystem changes in the early 1800s resulted from extensive hunting and trapping efforts led by the Russian-American Company, the Hudson's Bay Company, and others. Ranching, cattle grazing, and wheat farming were introduced in the mid-1800s on Mexican Ranchos such as Llano de Santa Rosa, Cabeza de Santa Rosa, and El Molino, and intensified as a result of accelerating population growth and settlement following the Gold Rush (Alley, Bowen & Co., Publishers 1880; Davis 1889; Gregory 1911; Sloop et al. 2007). An official report from the early 1860s stated that of the approximately 48,000 acres of cultivated land in Sonoma County at the time, a little over half was wheat and hay, with barley, oats, potatoes, and corn composing the majority of the remainder (*Sonoma County Democrat* 1863). Impacts from cattle ranching and grain farming during this era may have encouraged the conversion of native grasslands and forblands to grasslands dominated by non-native species and resulted in increased runoff and erosion from soil compaction (PWA 2004b). In addition, large numbers of oaks were girdled and felled to clear land for wheat cultivation and produce charcoal and firewood (Fig. 29; Taylor 1862, Waaland 1990, PWA 2004a).

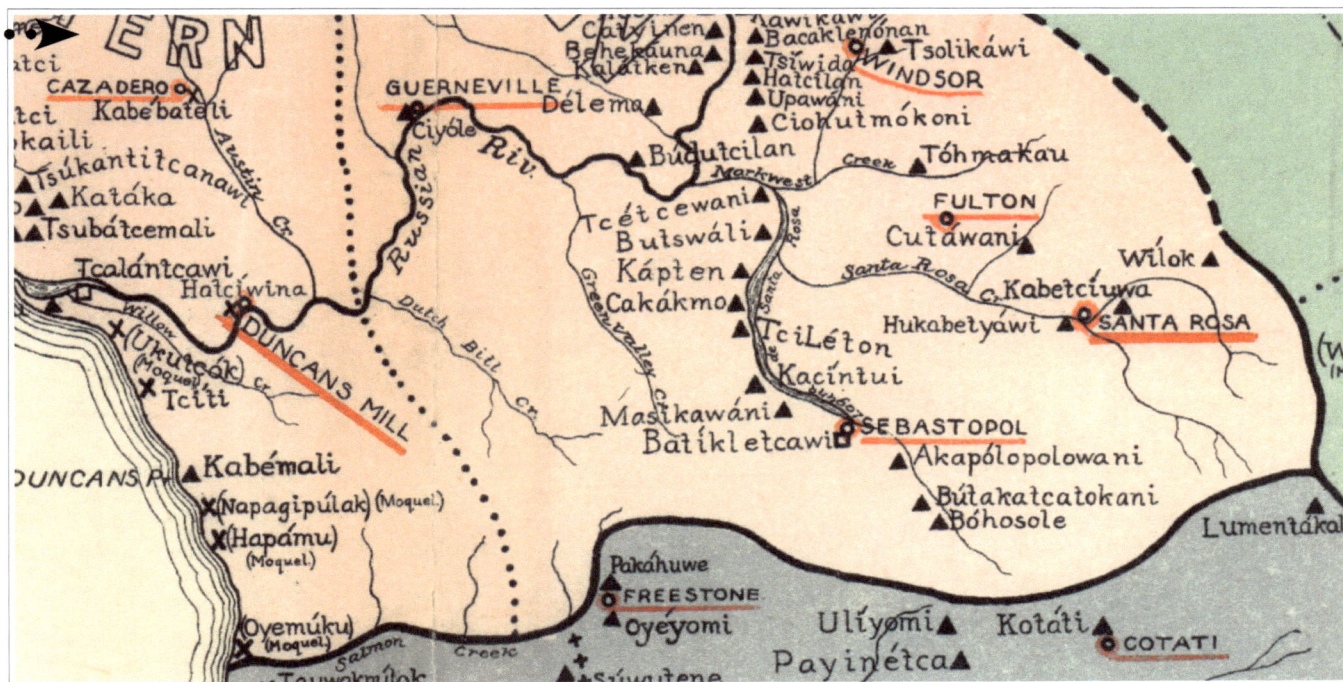

Fig. 27. A number of Pomo villages were situated around the Laguna de Santa Rosa, as illustrated by this 1908 linguistic map of Pomo territory. (Barrett 1908a, courtesy of The Bancroft Library, UC Berkeley)

Fig. 28. Pomo residents of Lake County. These images, taken by Edward Curtis in 1924, show Pomo inhabitants harvesting tules **(top)** and the use of tules for a variety of purposes such as constructing huts **(opposite)** and canoes **(bottom)**. Though in many cases representing "contrived reconstructions rather than true documentation" (Northwestern University Digital Library Collections 2003), Curtis' photographs nonetheless serve to illustrate some of the traditional land uses and cultural practices of the Pomo. *(Photo digital IDs: cp14008, cp14010, cp14018; courtesy of Charles Deering McCormick Library of Special Collections, Northwestern University)*

"As we got out of the shabby little village of Santa Rosa... It was melancholy to see how wantonly the most beautiful trees in the world had been destroyed; for the world has never seen such oaks as grow in the Russian River Valley. The fields of girdled and blackened skeletons seemed doubly hideous by contrast with the glory of the surviving trees."

– Taylor 1862

Fig. 29. (above) Hauling charcoal near the Laguna de Santa Rosa, ca. 1880. Large quantities of oaks were felled to clear land for ranching and farming and used to produce charcoal and firewood. *(Photo #2026, courtesy of Western Sonoma County Historical Society)*

The cultivation of orchards, vineyards, and row crops became widespread by the late 1800s, an endeavor that was made economical in large part by the arrival of the Northwestern Pacific Railroad in the 1870s (Watson et al. 1917). On a trip to the region in 1880, W.S. Walker described "immense stack-yards of Wheat, Oats and Barley; orchards groaning under their delicious burdens… Apples, Peaches, Pears, Plums, Figs, and Grapes abound, and Vegetables of every description are successfully grown" (Walker 1880). Near the Laguna, farmers took advantage of the fertile "bottomlands" and access to water to grow crops such as barley, oats, grapes, and hops (*Pacific Rural Press* 1874, *Pacific Rural Press* 1882); flood tolerant crops such as barley and oats occupied large areas of low elevation land that formerly supported seasonal wetlands (*Pacific Rural Press* 1880b, Watson et al. 1917, USDA 1942, Verhoeven and Setter 2010). Dairy production was also widespread along the margins of the Laguna and elsewhere on the Santa Rosa Plain in the late 19th and early 20th centuries (Fig. 30; *Pacific Rural Press* 1876, *Pacific Rural Press* 1880a, Watson et al. 1917, Cummings 2003a). Beginning in the 1940s and 1950s, much of the agricultural land in the region was developed to accommodate the region's increasing population (PWA 2004).

In addition to changes in land cover, each land-use era brought with it attendant changes to the hydrology and water resources of the Laguna. While irrigated agriculture was relatively uncommon in the region until the mid-20th century (Thompson 1884, Gregory 1911, Cummings 2003b), some surface diversions commenced as early as the late 19th century; for example, diversions from Santa Rosa Creek began in 1873 with the completion of Santa Rosa Water Works (Thompson 1877a). In contrast, by the mid-20th century, urban runoff and wastewater discharge were substantially augmenting dry season flows in the Laguna (Tancreto and River 1987, Waaland 1989, Cummings 2003b).

Fig. 30. (below) This ca. 1910 image shows the farm house of John A. and Barbara Ellen Brown, located on the eastern side of Sebastopol. The Laguna is visible in the background on the right side of the photo. The Browns owned 1,400 acres near the Laguna, which they leased to dairy farmers (Cummings 2003a). *(Photo #2045, courtesy of Western Sonoma County Historical Society)*

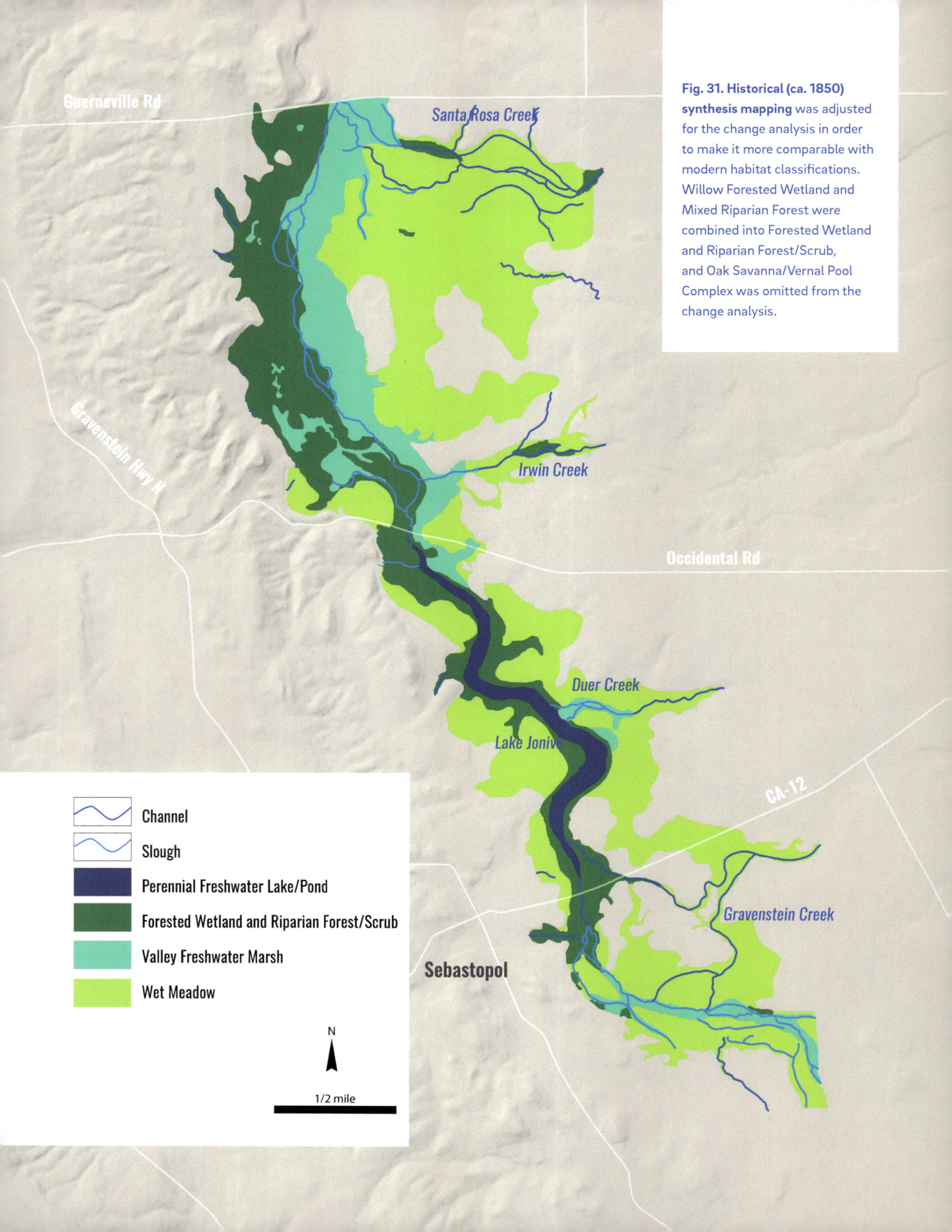

Fig. 31. Historical (ca. 1850) synthesis mapping was adjusted for the change analysis in order to make it more comparable with modern habitat classifications. Willow Forested Wetland and Mixed Riparian Forest were combined into Forested Wetland and Riparian Forest/Scrub, and Oak Savanna/Vernal Pool Complex was omitted from the change analysis.

Guerneville Rd

Santa Rosa Creek

Gravenstein Hwy N

Irwin Creek

Occidental Rd

Duer Creek

Lake Jonive

CA-12

Gravenstein Creek

Sebastopol

Channel
Slough
Perennial Freshwater Lake/Pond
Forested Wetland and Riparian Forest/Scrub
Valley Freshwater Marsh
Wet Meadow

N

1/2 mile

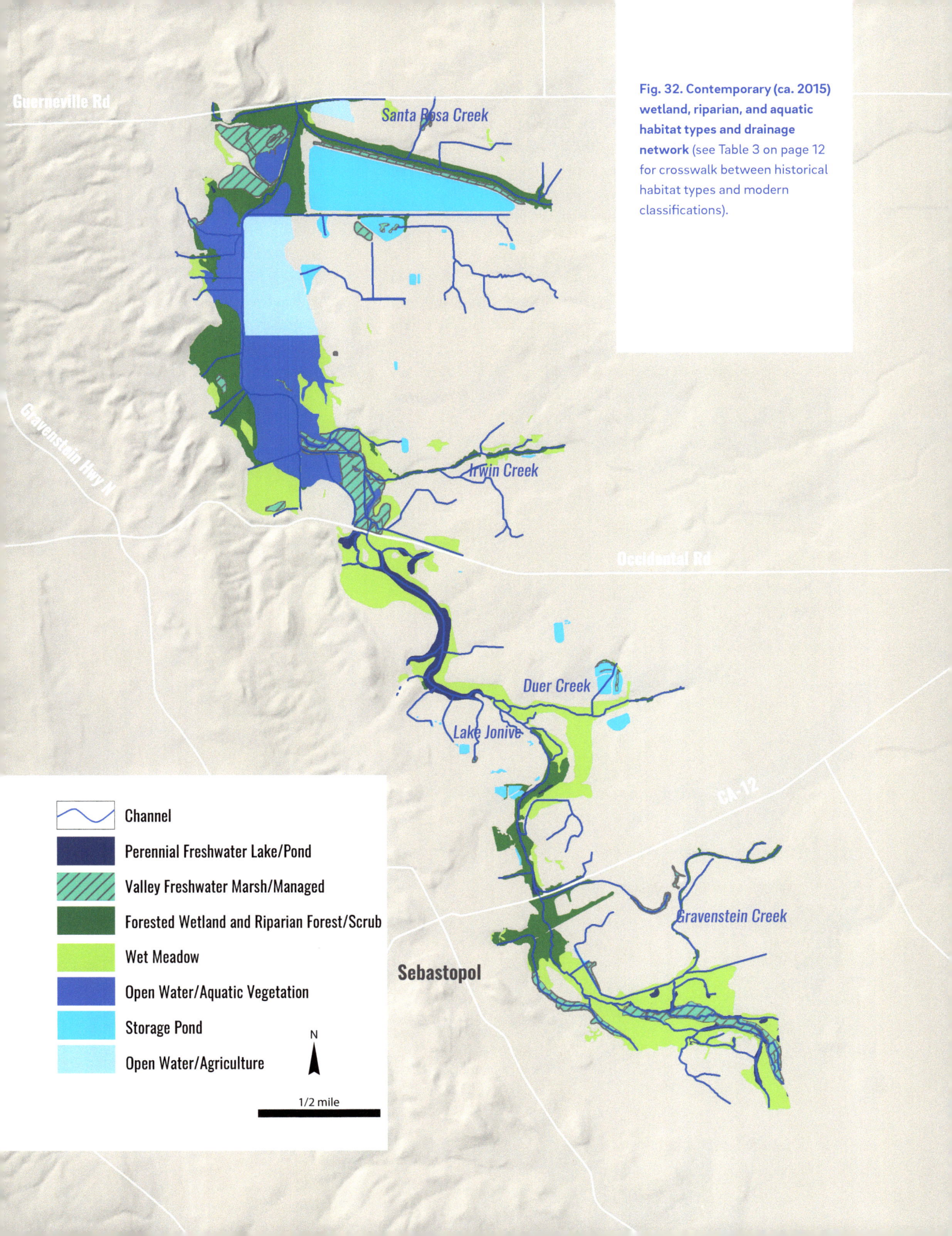

Fig. 32. Contemporary (ca. 2015) wetland, riparian, and aquatic habitat types and drainage network (see Table 3 on page 12 for crosswalk between historical habitat types and modern classifications).

Guerneville Rd

Santa Rosa Creek

Gravenstein Hwy N

Irwin Creek

Occidental Rd

Duer Creek

Lake Jonive

CA-12

Gravenstein Creek

Sebastopol

Legend

Channel

Perennial Freshwater Lake/Pond

Valley Freshwater Marsh/Managed

Forested Wetland and Riparian Forest/Scrub

Wet Meadow

Open Water/Aquatic Vegetation

Storage Pond

Open Water/Agriculture

N

1/2 mile

CHANGES IN HABITAT EXTENT AND DISTRIBUTION

Comparison of the historical (ca. 1850) and modern (ca. 2015) habitat type maps shows many of the dramatic changes that have occurred over the past century and a half (Figs. 31 and 32). These changes include habitat loss, conversion to other habitat types, and establishment of habitat types that did not exist historically. Fig. 33 summarizes changes in the extent of wetland, riparian, and aquatic habitat types within the study area.

Overall, there has been an approximately 47% loss of wetland, riparian, and aquatic habitat area within the study area (not including vernal pool complex). Among wetland and riparian habitat types only (excluding aquatic habitats), the loss is approximately 67%. All habitat types included in the change analysis have declined in area, including 26% loss of perennial freshwater lake/pond, 54% loss of forested wetland and riparian forest/scrub, 63% loss of valley freshwater marsh, and 73% loss of wet meadow. A visual inspection of the land cover data suggests that much of the lost habitat within the study area has been replaced with pastureland or cultivated areas. Declining groundwater levels in the late 20th century (Nishikawa et al. 2013) may have also contributed to wetland loss in some areas.

Fig. 33. Changes in extent of wetland, riparian, and aquatic habitat types. Overall, approximately half (47%) of total wetland, riparian, and aquatic habitat area has been lost within the study area.

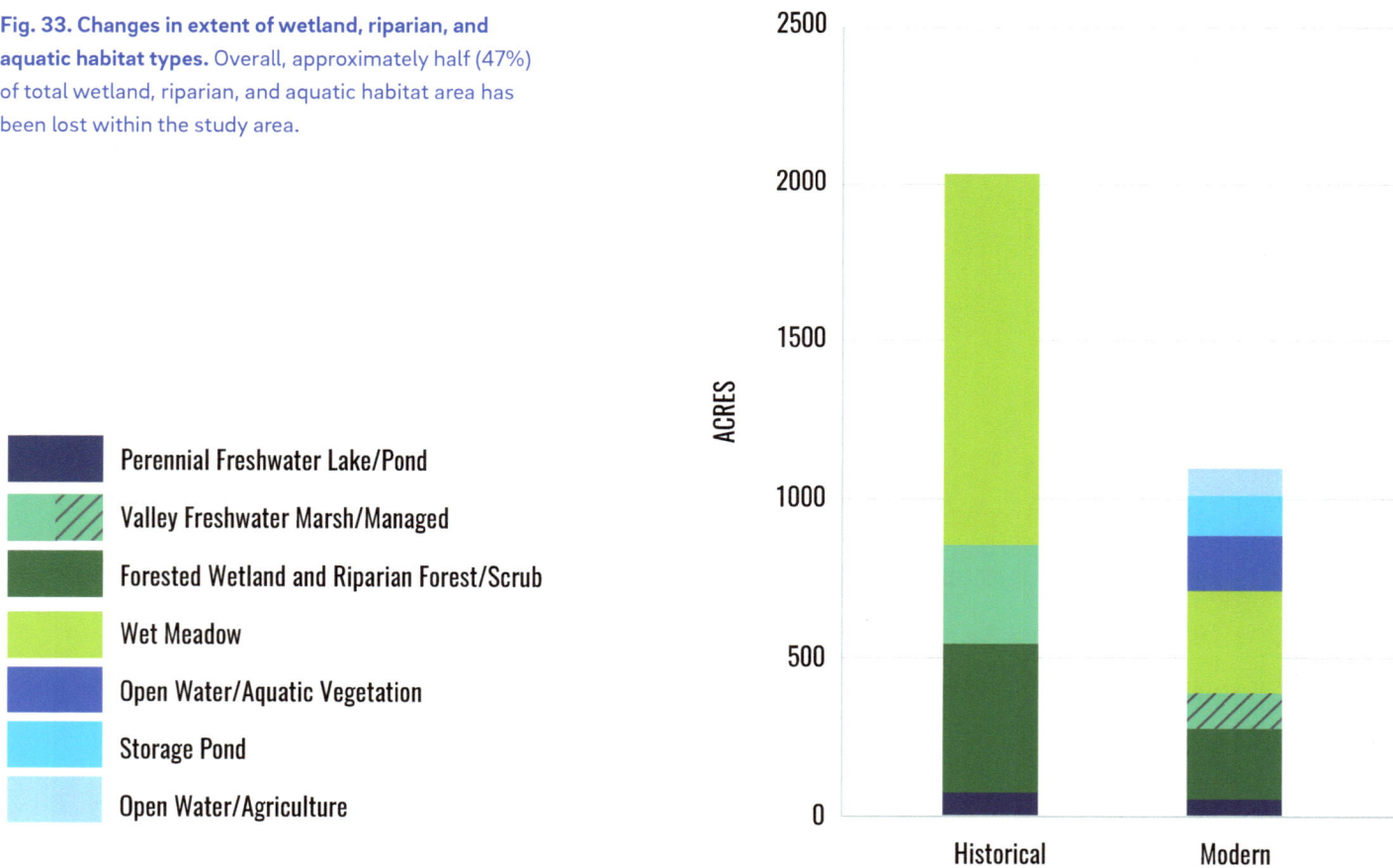

Legend:
- Perennial Freshwater Lake/Pond
- Valley Freshwater Marsh/Managed
- Forested Wetland and Riparian Forest/Scrub
- Wet Meadow
- Open Water/Aquatic Vegetation
- Storage Pond
- Open Water/Agriculture

One of the most notable changes is the loss of much of the broad riparian forest around Lake Jonive. Though a narrow riparian corridor still exists and provides valuable habitat and other ecological services, it is much narrower than the 10–350 m (30–1,150 ft) wide riparian buffer that historically surrounded the lake. The loss of riparian forest area would indicate that the remaining forest likely has a lower capacity to assimilate nutrients and other inputs from surrounding dairies and farms (see page 50). Lake Jonive itself has decreased in size by approximately 50% (from ~27 ha [66 ac] to ~14 ha [34 ac]), and is shallower today that it was during the mid-19th century (Butkus 2011a).

To the north of Lake Jonive, the large areas of valley freshwater marsh and willow forested wetland that existed historically have largely been converted to novel habitat types, including open water/aquatic vegetation (largely dominated by invasive *Ludwigia*) and open water/agriculture. This conversion has been driven in part by sediment accumulation: an analysis by PWA (2004b) estimated that up to 2 feet of sediment has accumulated between Occidental Road and the Santa Rosa Creek Flood Channel since the 1950s. Along with increased rates of sediment delivery from Santa Rosa Creek, Mark West Creek, and other tributaries (see below), dikes and berms constructed through the wetland restrict drainage and maintain shallowly flooded conditions throughout much of the year, favoring the establishment of aquatic vegetation over emergent marsh and forested wetland. Several small patches of emergent marsh, classified as Valley Freshwater Marsh/Managed, remain around the confluence with Irwin Creek and in the northwestern corner of the study area just south of Guerneville Road (a corridor of valley freshwater marsh/managed also remains in the southern part of the study area, around the confluence with Gravenstein Creek), but these marshes appear to be highly modified and likely do not provide the same ecological functions provided by the large freshwater marsh areas historically. Several patches of forested wetland have also persisted in the north and northwestern parts of the study area where the large willow forested wetland existed historically, but the hydrology and function of these wetlands has been significantly modified as well.

Though nearly three-quarters of its historical acreage has been lost, wet meadow has persisted in a number of areas throughout the study area, particularly in the south around the confluence with Gravenstein Creek, near the confluence of Duer Creek east of Lake Jonive, and around the Occidental Road crossing and the confluence with Irwin Creek. The largest patch of historical wet meadow, on the northeastern side of the study area, has been all but eliminated, as have large areas of wet meadow around Lake Jonive. The Delta Pond, a storage pond for treated wastewater, now occupies a large area of the floodplain near the confluence with Santa Rosa Creek that was historically dominated by wet meadow.

The loss and fragmentation of habitats in the Laguna has played a major role in the decrease in species support functions and biodiversity over time. Some species, such as Western yellow-billed cuckoo, have been extirpated from the region entirely, while others, such as coho salmon, California tiger salamander, and California freshwater shrimp, have been extirpated from portions of their historical range or have experienced severe declines in population size associated with loss of suitable habitat (Laguna Technical Advisory Committee 1989, USFWS 1998, Santa Rosa Plain Conservation Strategy Team 2005, Honton and Sears 2006, Sloop et al. 2007).

CHANGES IN THE CHANNEL NETWORK

In addition to changes in habitat extent and distribution, the configuration of the channel network within the study area has also been substantially modified over the past 150 years (Fig. 34).

Substantial portions of the channel network, including both the Laguna mainstem and tributaries, have been straightened and channelized in order to increase the efficiency of drainage and control flooding. These activities were carried out as part of the Central Sonoma Watershed Project (a joint effort between the Sonoma County Flood Control and Water Conservation District, the Santa Rosa Soil Conservation District, and the USDA Soil Conservation Service) which was created in 1958 in response to catastrophic flooding in the mid-20th century (Santa Rosa Soil Conservation District et al. 1958, Beach 2002, PWA 2004a). The existing flood control channel along Santa Rosa Creek was constructed during the 1960s (Fig. 35). Along the Laguna mainstem, an approximately 13 km (8 mi) reach from Occidental Road to a point about 1.6 km (1 mi) downstream of Guerneville Road was channelized in 1966 (Beach 2002, PWA 2004a, Cummings 2004). Where streams have been confined by artificial levees (such as along Santa Rosa Creek), the channels have been disconnected from their surrounding floodplains, altering hydrologic patterns and habitat conditions (Waaland 1989).

Several transportation corridors, including Guerneville Road, Occidental Road, Highway 12, and the Joe Rodota Trail (the former Petaluma and Santa Rosa Railway right-of-way) cross over the mainstem Laguna channel within the study area. Bridges on these corridors constrict the Laguna channel/floodplain and impede the movement of water and sediment downstream, likely resulting in increased sediment deposition within the Laguna channel and contributing to flooding upstream of the bridges.

The modifications to the channel network have a number of implications for water and sediment dynamics, nutrient transport, capacity to support wetland habitats, and aquatic habitat quality, among other factors. For example, sediment deposition from tributaries was historically dispersed throughout much of the wetland complex through a network of sloughs and distributary channels; today, in a number of areas, flows and sediment loads are concentrated in artificially straightened channels, leading to higher flow velocity and higher sediment input to the Laguna mainstem (Sloop et al. 2007, TetraTech 2015b). Increased sediment deposition within the study area, and particularly at the mouth of Santa Rosa Creek, has resulted in decreased flood storage capacity and contributed to the conversion to aquatic habitat (PWA 2004b, Curtis et al. 2013, TetraTech 2015b).

The decrease in flood storage (and hence the increase in flood risk) has been exacerbated by changes in the alignment of Mark West Creek, which connects with the Laguna mainstem approximately 1 km (0.7 mi) north of Guerneville Road (PWA 2004a, Winzler & Kelly-GHD 2012). Historically, lower Mark West Creek flowed northwest to connect with the Laguna approximately 0.75 km (0.5 mi) north of present-day River Road. Over the past 150 years, a series of channel modifications shifted the Laguna-Mark West Creek confluence approximately 3.2 km (2 mi) south. As a result, much of the sediment brought down by Mark West Creek during high flows is deposited in or near the Laguna just north of the study area, which contributes to flooding and other problems upstream (Fig. 36; Baumgarten et al. 2014a).

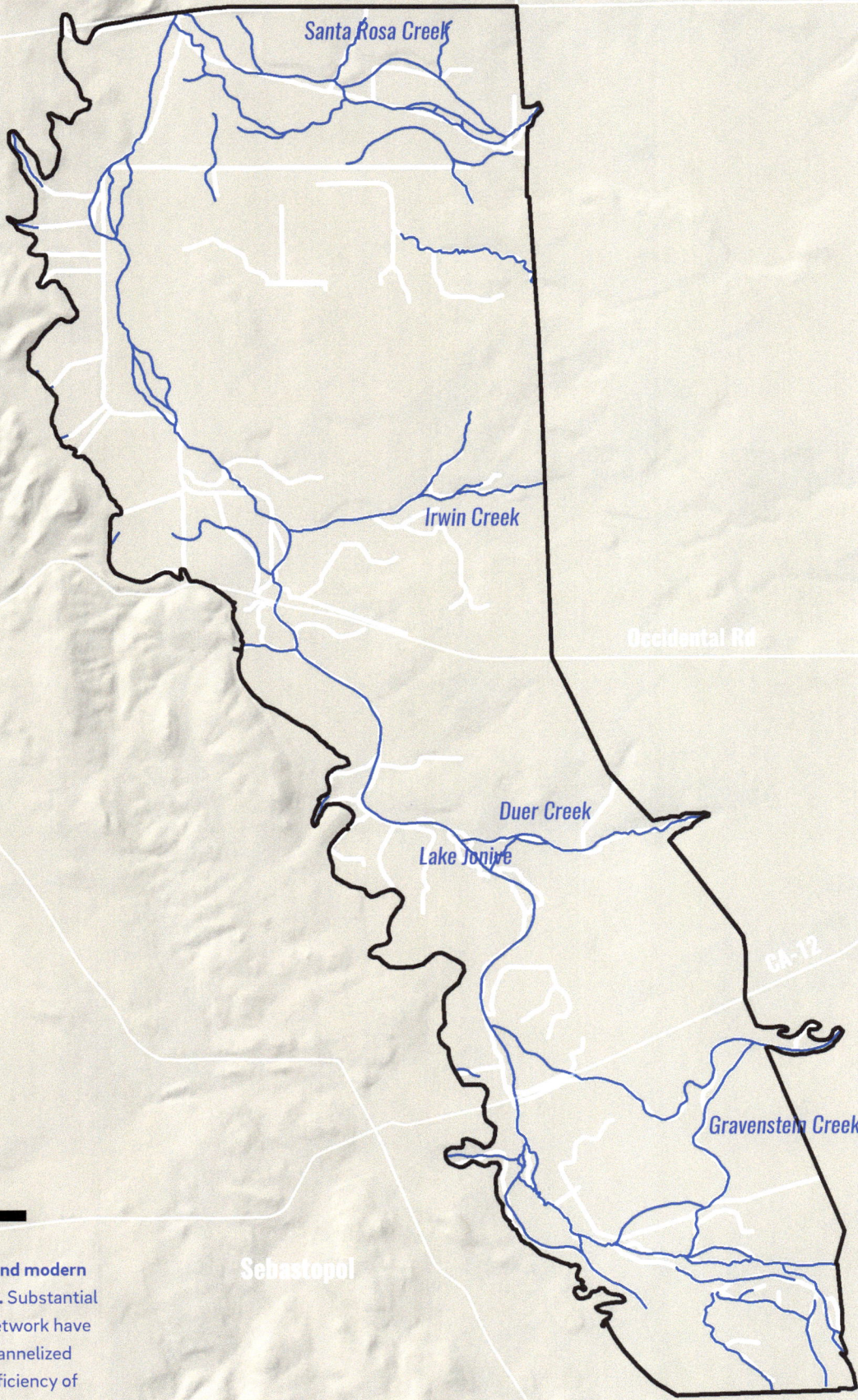

Santa Rosa Creek

Gravenstein Hwy R

Guerneville Rd

Irwin Creek

Occidental Rd

Duer Creek

Lake Jonive

CA-12

Gravenstein Creek

N

1/2 mile

Sebastopol

Fig. 34. Historical (blue) and modern (white) channel networks. Substantial portions of the channel network have been straightened and channelized in order to increase the efficiency of drainage and control flooding.

Fig. 35. Construction of the Santa Rosa Creek Flood Control Channel (bottom) in 1964, and completed channel (top) in 1967 looking west towards Laguna. (Top: Print no. 0-3, Location = "Wendell Prop"; Bottom: Photo #999, "Unit No. 2 Santa Rosa Creek Channel Lower Channel 3 construction"; both courtesy of Sonoma County Water Agency)

Fig. 36. (opposite) Change over time in the alignment of Mark West Creek, 1850-2012, north of the study area. A series of shifts intended to control flooding and sediment deposition have moved Mark West Creek's confluence with the Laguna approximately 3.2 km (2 mi) south of its historical location. (from Baumgarten et al. 2014a)

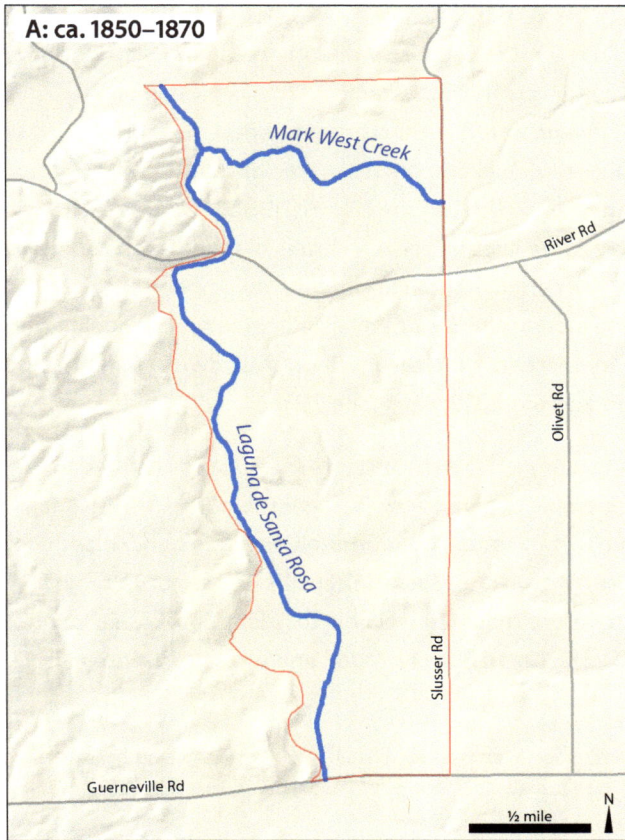

A: ca. 1850–1870

Mark West Creek

River Rd

Olivet Rd

Laguna de Santa Rosa

Slusser Rd

Guerneville Rd

½ mile

N

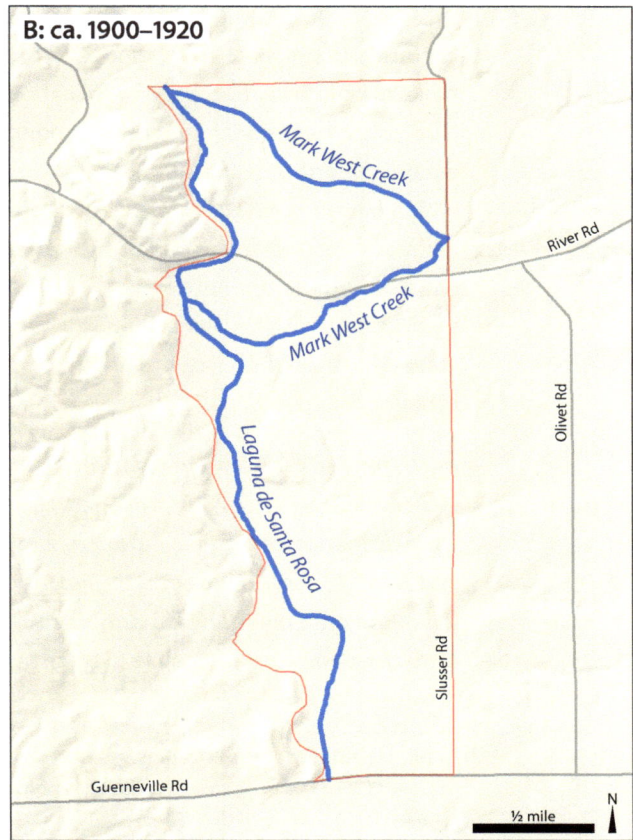

B: ca. 1900–1920

Mark West Creek

Mark West Creek

River Rd

Olivet Rd

Laguna de Santa Rosa

Slusser Rd

Guerneville Rd

½ mile

N

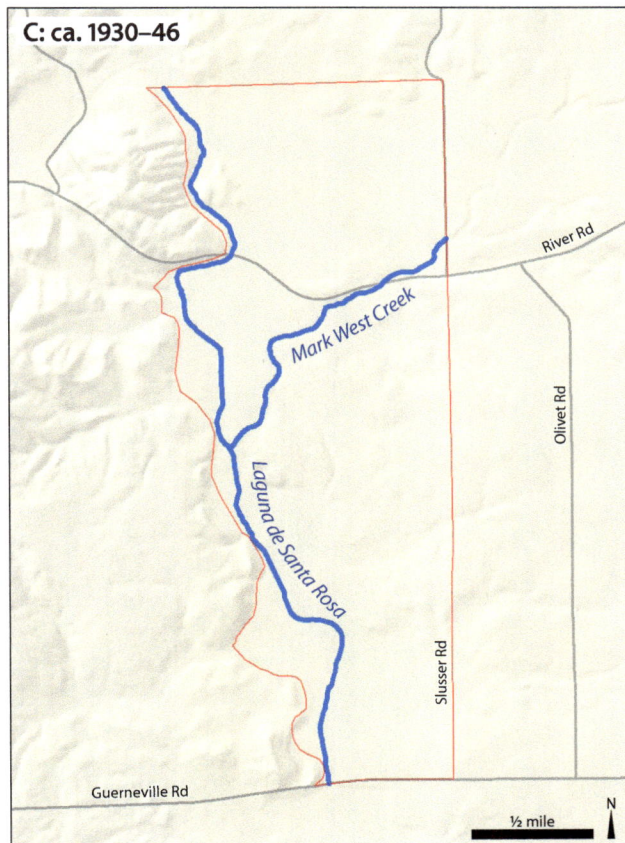

C: ca. 1930–46

River Rd

Mark West Creek

Olivet Rd

Laguna de Santa Rosa

Slusser Rd

Guerneville Rd

½ mile

N

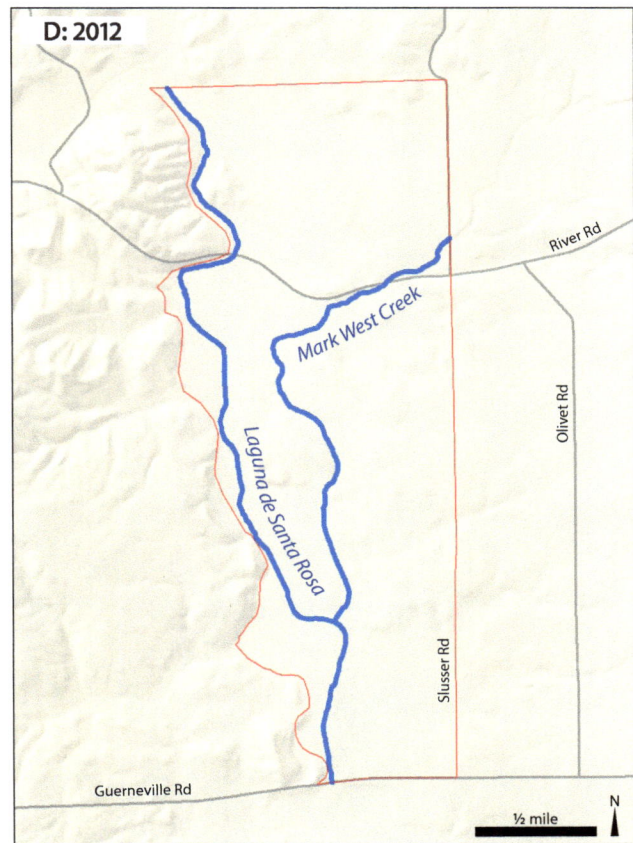

D: 2012

River Rd

Mark West Creek

Olivet Rd

Laguna de Santa Rosa

Slusser Rd

Guerneville Rd

½ mile

N

CHANGES IN ASSIMILATIVE CAPACITY

Historical changes to the drainage network, along with changes in wetland extent and distribution, have cumulatively had major implications for water quality in the Laguna. This is particularly true given the broader land use changes that have occurred within the watershed, such as the establishment of dairies and other agricultural enterprises and the discharge of treated wastewater, which have increased external nutrient loading rates to the Laguna (Morris 1995).[1] Butkus (2013) estimates that nitrogen and phosphorus inputs to the Laguna are 4–6 times higher today than they were during the pre-settlement era. Increased nutrient loading has significantly impacted habitat quality, and, along with hydrologic changes, the increased nutrient availability (and in particular, the elevated levels of phosphorus) is likely a primary factor driving the invasion by *Ludwigia* (Fig. 37) and other non-native aquatic plants (Baye 2008, Fitzgerald 2013, TetraTech 2015a).

As water and nutrients move through a wetland, some of those nutrients may be sequestered in the soil or released into the atmosphere. Under anaerobic conditions (which are the norm in wetlands), transformation of organic nitrogen, ammonium, and nitrate may ultimately result in denitrification and loss of gaseous nitrogen to the atmosphere (Mitsch and Gosselink 1993). Phosphorus can be adsorbed onto clays and other minerals and therefore made unavailable for plant uptake, although these reactions are reversible (Sanyal and De Datta 1991). Both nitrogen and phosphorus are also

---

[1] Internal nutrient sources, representing legacy contamination from past land uses, also contribute significantly to nutrient loads to the Laguna, but are not addressed in this analysis (TetraTech 2015a, D. Kuszmar pers. comm.).

Fig. 37. The spread of invasive Ludwigia over large areas of the Laguna has been driven in part by increased nutrient loads, along with changes in hydrology and sediment dynamics (Tetratech 2015a). Ludwigia is visible in the Laguna between Occidental Road and Guerneville Road in this 2006 photograph. *(Photo by Julian Meisler, 2006, courtesy of Laguna de Santa Rosa Foundation)*

assimilated into plant biomass, though for the most part this only immobilizes them temporarily (a small fraction of the nutrients stored in plant biomass may be immobilized more permanently through the accumulation of dead vegetation and development of organic soils; Nichols 1983).

The assimilative capacity of a wetland (its ability to remove or filter nutrient loads) depends on a number of factors, including wetland type. Because they maintain anaerobic conditions throughout the year, perennial wetlands are generally understood to remove a greater proportion of nitrogen than seasonal wetlands. However, anaerobic conditions increase the solubility of phosphorus, and can thus reduce phosphorus retention within the soil (Uusi-Kämppä et al. 2001). Thus, different types of wetlands may be maximally effective at assimilating either nitrogen or phosphorus: wetlands that are permanently flooded or have a perennially high water table (such as valley freshwater marsh) will tend to promote higher rates of denitrification, while drier wetlands or riparian habitat types that are only temporarily or seasonally flooded likely promote greater phosphorus retention (Haycock et al. 2001, Fisher and Acreman 2004, Hefting et al. 2004).

In general, the larger the area of the wetland, and the slower the water moves through it (i.e., the greater the residence time), the greater the proportion of nutrients that will be removed (Richardson and Qian 1999, Knox et al. 2008). In systems where floodplains are intact and connected to the channels, water and nutrients will spread out across the floodplain, slowing flow velocities and increasing residence time (Knox et al. 2008). In addition, the rate at which nutrients are removed depends on the concentration of nutrient inputs (the efficiency of removal tends to be higher for lower nutrient loading rates), soil type (e.g., soils with high concentrations of organic matter may have greater capacity to remove nitrogen), soil pH (the rate of denitrification is more rapid in neutral or alkaline soils), temperature, and season (Nichols 1983, Mitsch and Gosselink 1993, Fisher and Acreman 2004).

Figs. 38 and 39 present a conceptual model of how the assimilative capacity of the central Laguna wetland complex has changed over the past 150 years, as well as potential changes in the future resulting from wetland restoration. Historically, there was a relatively small nutrient load entering the system.[2] This load moved slowly through large wetland complexes, which presumably removed a large portion of the nutrients, resulting in a very small load exiting the system. Today, external nutrient inputs are much higher than they were historically (resident/internal nutrient loads are also significant), and there is much less wetland area to remove nutrients. Channelization of portions of the Laguna mainstem and tributaries has decreased floodplain connectivity and the degree of channel-wetland interaction, and has also likely resulted in more rapid flow through the system in some areas (shorter residence time). As a result, the assimilative capacity of the system is much lower today than it was historically, and there is a much higher nutrient load exiting the system (Butkus 2013). Although a restoration feasibility analysis was not conducted as part of this study, examination of the historical wetland distribution, as well as preliminary exploration of current hydrologic conditions and land uses within the study area, suggests that there may be potential for wetland restoration to increase assimilative capacity in a number of locations.

---

[2] While the historical nutrient load was small compared with modern loading rates, the Laguna was likely always a somewhat eutrophic (nutrient-rich) system. This was due in large part to its low gradient and the resulting low surface flow velocity, which allowed a large amount of nutrient-rich sediment and organic matter to settle out on the floodplain. The low flow velocity also resulted in longer residence times, however, allowing for greater nutrient assimilation and potentially offsetting some of the effects of the high organic matter input (Waaland 1990).

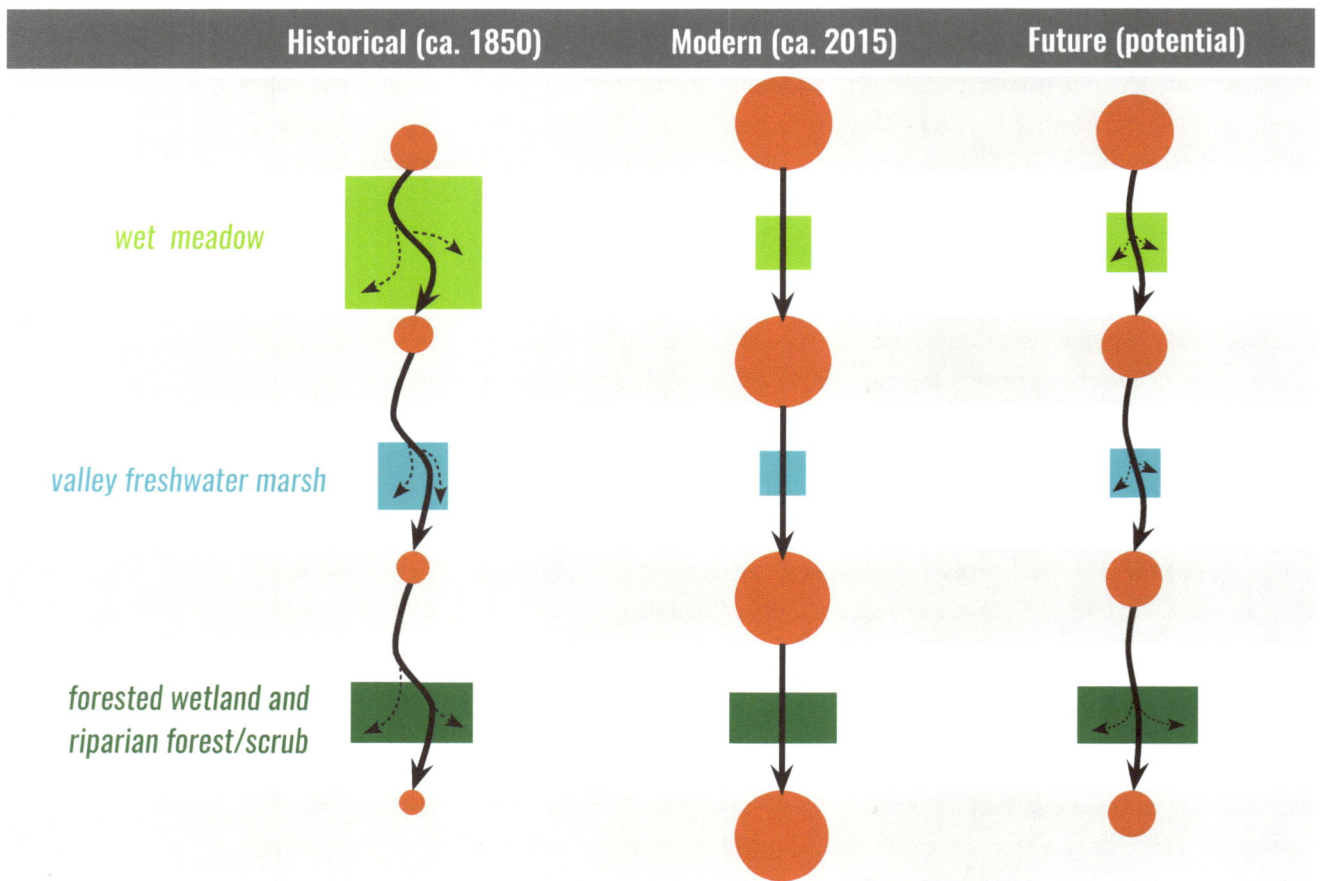

Fig. 38. Conceptual model representing past and potential future changes in nutrient assimilative capacity within the study area. Red circles represent external nitrogen and phosphorus loads as they move through the system. The colored boxes represent the wetland and riparian habitat types most likely to contribute significantly to nutrient assimilation. There is potential to mitigate external nutrient loading to the central Laguna in the future, both by reducing nutrient inputs from sources, and by restoring wetlands in strategic locations to increase the assimilative capacity of the system.

Photo by Sean Baumgarten,
January 13, 2017

Controls	Historical (ca. 1850)	Modern (ca. 2015)	Future (potential)
Nutrient inputs			
Wetland area			
Floodplain connectivity			
Residence time			
Wetland interaction			

Fig. 39. Conceptual summary of the key controls believed to be responsible for changes in assimilative capacity over time within the study area. Relative to modern conditions, the historical Laguna had lower nutrient inputs, greater wetland area, more floodplain connectivity, and greater residence time, all of which contributed to a higher assimilative capacity of the system and lower nutrient export from the system.

SYNTHESIS AND NEXT STEPS

The transformations to the central Laguna de Santa Rosa, and to the broader Laguna watershed, over the past 150 years have had far-reaching impacts on the physical and ecological functioning of the system. Where historically the central Laguna was an expanse of diverse wetland, riparian, and aquatic habitat types arranged along gradual topographic and hydrologic gradients, today the landscape is dominated by a highly managed patchwork of habitat types representing varying degrees of modification. While patches of relatively intact wetland habitats persist in many parts of the study area, the extent of all native habitat types has declined, and novel habitats dominated by invasive aquatic species now occupy large portions of the wetland complex. Habitat loss has been accompanied by extensive channelization and other changes to the stream network, which have substantially altered streamflow and sediment dynamics. The loss of native wetland habitats, along with changes to the channel network and increased nutrient inputs from the watershed, have likely substantially lowered the assimilative capacity of the system and contributed to elevated nutrient loading to surface waters.

The analysis presented here suggests that there is the potential to address excess nutrient loading in the Laguna in part through strategic restoration of native habitat types such as valley freshwater marsh, wet meadow, forested wetland, and riparian forest. Different wetland and riparian habitat types vary in their effectiveness at removing different nutrients, and thus a restoration plan aimed at reducing multiple forms of both nitrogen and phosphorus and maximizing total assimilative capacity should incorporate a portfolio of different habitats. Perennially flooded wetland types that maintain permanent anaerobic conditions would likely be more effective at removing nitrogen from the system, while habitat types that are temporarily or seasonally flooded, such as riparian forests, may be more appropriate for phosphorus assimilation.

The strategic restoration of wetland and riparian habitats within the Laguna basin, along with restoration of more natural hydrologic processes, could provide a number of additional benefits beyond water quality improvement: co-benefits may include support for native plants and wildlife, groundwater recharge, sediment storage, flood attenuation, and other desired outcomes. For example, in areas where the Laguna mainstem has been channelized (e.g., north of Occidental Road), restoration of surrounding wetland areas and reconnection of the Laguna channel with its floodplain would likely reduce in-channel sediment deposition and increase flood storage capacity. Where roads or other infrastructure artificially constrict water and sediment movement (e.g., at the Highway 12, Occidental Road, and Guerneville Road bridges), redesign of these barriers would improve hydrologic functioning and help to alleviate flooding and sedimentation problems. Removal of barriers and re-establishment of natural corridors connecting currently fragmented habitats would also facilitate wildlife movement and enhance the ecological resilience of the system to both natural and anthropogenic disturbances and stressors.

Site-specific analysis will be needed to examine the feasibility of particular restoration opportunities in relation to hydrology, substrate, land ownership, and other factors. However, many areas within the study area that historically supported wetlands are currently undeveloped and are in public ownership. In addition, preliminary examination of current hydrologic conditions (e.g., groundwater levels, drainage connectivity) suggests that there may be opportunities for wetland restoration in a number of areas. For example, following declines during the 1970s–90s, groundwater levels within the Cotati Basin Storage Unit have substantially recovered (Nishikawa et al. 2013), and many areas within the central portion of the Laguna still receive groundwater input (CWMW 2016).

SFEI's historical ecology research within the Laguna watershed to date has been limited in geographic scope primarily to portions of the central and lower Laguna and surrounding wetland areas. Future research will build on this existing work and provide for a more holistic understanding of historical ecological patterns and landscape functioning throughout the Laguna. Beginning in mid-2017, SFEI, in collaboration with the Sonoma County Water Agency and the Laguna de Santa Rosa Foundation, and with funding from the California Department of Fish and Wildlife, will develop a historical reconstruction and analysis of landscape change for the full Laguna as a foundation for the development of a Master Restoration Plan for the Laguna watershed, anticipated to be completed in 2020. The Restoration Plan will provide a vision and a framework for future restoration efforts in the Laguna, and will prioritize steps that can be taken to improve the health of the watershed and restore lost or degraded ecological functions.

ABAG (Association of Bay Area Governments). 2005. *Existing land use 2005: Sonoma County.* Oakland, CA.

Alley, Bowen & Co., Publishers. 1880. *History of Sonoma County, including its geology, topography, mountains, valleys and streams; together with a full and particular record of the Spanish grants; its early history and settlement, compiled from the most authentic sources; the names of original Spanish and American pioneers; a full political history, comprising the tabular statements of elections and office-holders since the formation of the county; separate histories of each township, showing the advancement of grape and grain growing interests, and pisciculture; also, incidents of pioneer life; the raising of the bear flag; and biographical sketches of early and prominent settlers and representative men; and of its cities, towns, churches, schools, secret societies, etc. etc.* San Francisco: Alley, Bowen & Co., Publishers.

Anderson MK. 2005. *Tending the wild.* Berkeley, CA: University of California Press.

Barrett SA. 1908a. Map of the territory of the Pomo linguistic stock and of the adjacent territories of other linguistic stocks, showing dialectic subdivisions and village sites. Berkeley: The University Press. *Courtesy of The Bancroft Library, UC Berkeley.*

Barrett SA. 1908b. The ethno-geography of the Pomo and neighboring Indians. *University of California publications in American archaeology and ethnology 6(1).* Berkeley: The University Press. *Courtesy of The Bancroft Library, UC Berkeley.*

Baumgarten S, Beller E, Grossinger R, et al. 2014a. *Historical changes in channel alignment along Lower Laguna De Santa Rosa and Mark Creek.* San Francisco Estuary Institute, Richmond, CA.

Baumgarten S, Striplen C, Beller E, et al. 2014b. *Historical ecology reconnaissance of the Middle Reach Russian River.* San Francisco Estuary Institute, Richmond, CA.

Baumgarten S, Striplen C, Salomon M, et al. 2014c. *Santa Rosa Plain orthoimagery and historical ecology study.* San Francisco Estuary Institute, Richmond, CA.

Baye P. 2008. *Summary of field trip discussion topics – Laguna de Santa Rosa floodplain wetlands, Graton, July 15, 2008.*

Beach RF. 2002. *History of the development of the water resources of the Russian River.* Sonoma County Water Agency.

Bowers AB. 1866. Map of Sonoma County, California. *Courtesy of Rumsey Map Collection.*

Bowers AB. 1867. Map of Sonoma County. *Courtesy of Sonoma State University North Bay Regional & Special Collections.*

Butkus S. 2011a. *Constructing stream flow rating power equations for the pre-settlement lakes in the Laguna de Santa Rosa watershed.* North Coast Regional Water Quality Control Board. Santa Rosa, CA.

Butkus S. 2011b. *Development of the Laguna de Santa Rosa watershed pre-European settlement spatial data model.* North Coast Regional Water Quality Control Board. Santa Rosa, CA.

REFERENCES

Butkus S. 2013. *Development of the land cover loading model for the Laguna de Santa Rosa Watershed*. California Regional Water Quality Control Board, North Coast Region.

California State Earthquake Investigation Commission. 1908. Map of the city of Santa Rosa, Sonoma County, California showing the portions destroyed by the earthquake of April 18, 1906, and by the fire consequent thereto. Baltimore: A. Hoen & Co. *Courtesy of David Rumsey Map Collection.*

Cardwell G. 1958. Geological survey water-supply paper 1427. *Geology and ground water in the Santa Rosa and Petaluma Valley areas, Sonoma County California*. U.S. Government Printing Office.

Cummings J. 2003a. *"Crystal Laughing Waters" - historical glimpses of the Laguna de Santa Rosa*. Petaluma, CA. *Courtesy of Sonoma State University Library North Bay Digital Collections.*

Cummings J. 2003b. *The awful offal of Sebastopol*. Petaluma, CA. *Courtesy of Sonoma State University Library North Bay Digital Collections.*

Cummings J. 2004. *Draining and filling the Laguna de Santa Rosa*. Petaluma, CA. *Courtesy of Sonoma State University Library North Bay Digital Collections.*

Cummings J. 2006. *Early Sebastopol: part IV - "sprightly Sebastopol" - a "lively burg"*. Petaluma, CA. *Courtesy of Sonoma State University Library North Bay Digital Collections.*

Curtis JA, Flint LE, Hupp CR. 2013. Estimating floodplain sedimentation in the Laguna de Santa Rosa, Sonoma County, CA. *Wetlands* 33(1):29-45.

CWMW (California Wetlands Monitoring Workgroup). 2016. California rapid assessment method for wetlands and riparian areas (CRAM). *EcoAtlas*. http://www.ecoatlas.org/regions/ecoregion/statewide?cram=1. Accessed September 22, 2016.

Daily Alta California. 1866. The storm in the interior. December 22. *Courtesy of California Digital Newspaper Collection.*

Daily Alta California. 1872. Sonoma County letter no. one: its towns, scenery, climate, natural resources - Petaluma, Santa Rosa, their promising future. October 30. *Courtesy of California Digital Newspaper Collection.*

Daily Alta California. 1888. A city's sewage. October 23. *Courtesy of California Digital Newspaper Collection.*

Davis P. 1879. Plat of the Laguna Drainage District [Recorder map no. 07m14]. *Courtesy of Sonoma County Recorder.*

Davis WH. 1889. Sixty years in California: a history of events and life in California; personal, political and military, under the Mexican regime; during the quasi-military government of the territory by the United States, and after the admission of the state into the Union, being a compilation by a witness of the events described. San Francisco: A.J. Leary, Publisher.

Delattre MP, Koehler R. 2008. Geologic map of the Sebastopol 7.5' quadrangle Sonoma County, California: a digital database. California Department of Conservation.

De Mars J, Johnson M, Lipshutz T, et al. 1977. *Laguna de Santa Rosa environmental analysis and management plan. Courtesy of Sonoma County Library.*

Denner S. 2002. *Transcript of oral interview of Stan Denner recorded by Jane Nielson on 8/30/02.*

Dyer EH. 1861. Plat of the Llano de Santa Rosa, finally confirmed to Joaquin Carillo. Surveyed under instructions from the U.S. Surveyor General. *Courtesy of Bureau of Land Management.*

Figueroa J. 1834. *Diario de la expedicion al otro lado de la Bahia de San Francisco por el general Figueroa.* José Figueroa papers, 1833-1835, BANC MSS C-A 239. *Courtesy of The Bancroft Library, UC Berkeley.*

Fisher J, Acreman MC. 2004. Wetland nutrient removal: a review of the evidence. *Hydrology and Earth System Sciences* 8(4): 673-685.

Fitzgerald R. 2013. *Summary of TMDL development data pertaining to nutrient impairments in the Laguna de Santa Rosa watershed (revised).* North Coast Regional Water Quality Control Board. Santa Rosa, CA.

Fredrickson DA, Markwyn DW. 1990. Cultural resources of the Laguna. In *History, land uses and natural resources of the Laguna de Santa Rosa*, ed. David W. Smith Consulting, 3-1 to 3-12. *Courtesy of Sonoma County History & Genealogy Library.*

Gray N. 1857. *Field notes of the final survey of the Rancho los Molinos, John B.R. Cooper, confirmee.* U.S. Department of the Interior, Bureau of Land Management. Book C-25/G3. *Courtesy of Bureau of Land Management.*

Gregory T. 1911. *History of Sonoma County, California, with biographical sketches of the leading men and women of the County, who have been identified with its growth and development from the early days to the present time.* Los Angeles: Historic Record Company.

Grossinger RM, Striplen CJ, Askevold RA, et al. 2007. Historical landscape ecology of an urbanized California valley: wetlands and woodlands in the Santa Clara Valley. *Landscape Ecology* 22:103-120.

Haycock NE, Pinay G, Burt TP, Goulding KWT. 2001. Buffer zones: current concerns and future directions. In *Buffer Zones: Their Processes and Potential in Water Protection, The Proceedings of the International Conference on Buffer Zones, September 1996*, ed. Nick Haycock, Tim Burt, Keith Goulding, and Gilles Pinay. Haycock Associated Limited.

Hefting M, Clément JC, Dowrick D, Cosandey AC, Bernal S, Cimpian C, Tatur A, Burt TP, Pinay G. 2004. Water table elevation controls on soil nitrogen cycling in riparian wetlands along a European climatic gradient. *Biogeochemistry* 67: 113-134.

Heintz MB. 1990. *Ballard Lake.* Gaye LeBaron Collection, Box 043. *Courtesy of Sonoma State University Library Special Collections.*

Holway RS. 1913. *The Russian River: A characteristic stream of the California coast ranges.* University of California Publications in Geography 1(1):1-60.

Honton J, Sears AW. 2006. *Enhancing and caring for the Laguna.* Santa Rosa, CA: Laguna de Santa Rosa Foundation.

Howell JT. 1951. Consortium of California Herbaria, record for *Fraxinus latifolia* from "Laguna east of Sebastopol." California Academy of Sciences.

Knox AK, Dahlgren RA, Tate KW, Atwill ER. 2008. Efficacy of natural wetlands to retain nutrient, sediment, and microbial pollutants. *Journal of Environmental Quality* 37: 1837-1846.

Kroeber AL. 1925. *Handbook of the Indians of California.* New York: Dover Publications, Inc.

Laguna Technical Advisory Committee. 1989. *Fish and wildlife restoration of the Laguna de Santa Rosa, Sonoma County, California. Courtesy of Sonoma County History & Genealogy Library.*

Lanman CW, Lundquist K, Perryman H, et al. 2013. The historical range of beaver (*Castor canadensis*) in coastal California: an updated view of the evidence. *California Fish and Game* 99(4):193-221.

Lee, Charles H. 1944. *Sonoma County sanitary survey: report on sanitary survey of Sonoma County, California with recommendations for control of epidemics.* [Unpublished] Report for the Sonoma County Board of Supervisors and the Sonoma County Department of Public Health. San Francisco, CA. 110pp.

The Lewis Publishing Company. 1889. *An illustrated history of Sonoma County, California. Containing a history of the County of Sonoma from the earliest period of its occupancy to the present time, together with glimpses of its prospective future: with profuse illustrations of its beautiful scenery, full-page portraits of some of its most eminent men, and biographical mention of many of its pioneers and also of prominent citizens of today.* Chicago, IL: The Lewis Publishing Company.

Marryat F. 1855. *Mountains and molehills or recollections of a burnt journal.* London: Longman, Brown, Green, and Longmans.

Martin HB, Eliason MA. 1861a. *Survey no. 14. Township 6 North Range 9 West. Section 1. Fraction area 191 1/2 acres. Mt. D. Meridian. October 12.* Official Surveys Vol. 1 1850-1873, p. 238. Sonoma County Surveyor, Permit and Resource Management Department. *Courtesy of Sonoma County Surveyor.*

Martin HB, Eliason MA. 1861b. *Survey no. 41. Township 7 North Range 9 West. Section 22. November 6.* Official Surveys Vol. 1 1850-1873, p. 265. Sonoma County Surveyor, Permit and Resource Management Department. *Courtesy of Sonoma County Surveyor.*

Menefee CA. 1873. *Historical and descriptive sketch book of Napa, Sonoma, Lake and Mendocino, comprising sketches of their topography, productions, history, scenery, and peculiar attractions.* Napa City: Reporter Publishing House.

Merriam CH. 1966. Ethnographic notes on California Indian tribes. In *Reports of the University of California Archaeological Survey, No. 68, Part I,* ed. Robert F. Heizer. University of California Archaeological Research Facility, Department of Anthropology.

Millington S. 1865. *Transcript of the field notes of the survey of the subdivision lines Township 7 North Range 9 West, Mount Diablo Meridian, State of California.* Book 182-9. U.S. Department of the Interior, Bureau of Land Management. *Courtesy of Bureau of Land Management.*

Mitsch WJ, Gosselink JC. 1993. *Wetlands.* Second edition. New York: Van Nostrand Reinhold.

Moraga G. 1810. *Diario de su expedicion al Puerto de Bodega.* Provincial State Papers Tom. XIX 1805-1815, p. 278. BANC MSS C-A 12. https://ia802607.us.archive.org/5/items/168036075_79_14/168036075_79_14.pdf. Accessed March 13, 2017. *Courtesy of The Bancroft Library, UC Berkeley.*

Morris CN. 1995. *Waste reduction strategy for the Laguna de Santa Rosa*. North Coast Regional Water Quality Control Board, Santa Rosa, CA.

Mueller CE. 1921. Jewell's subdivision of the Bisordi Bro's Allendale Ranch, being portions of sections 26 and 27 of Township 7 North, Range 9 West M.D.M., Sonoma County, California [Recorder map no. 40m06]. *Courtesy of Sonoma County Recorder.*

NCRWQCB (North Coast Regional Water Quality Control Board). 2011. *Water quality control plan for the North Coast region*. North Coast Regional Water Quality Control Board, Santa Rosa, CA.

Nichols DS. 1983. Capacity of natural wetlands to remove nutrients from wastewater. *Journal (Water Pollution Control Federation)* 55(5): 495-505.

Nishikawa T, Hevesi JA, Sweetkind DS, et al. 2013. *Hydrologic and geochemical characterization of the Santa Rosa Plain watershed, Sonoma County, California: Chapter A: Introduction to study area.* Sonoma County, California.

Northwestern University Digital Library Collection. 2003. *About the work.* Edward S. Curtis's The North American Indian. http://curtis.library.northwestern.edu/aboutwork.html. Accessed March 25, 2017.

Origer RM, Fredrickson DA. 1980. *The Laguna Archaeological Research Project Sonoma County, California*. Cultural Resources Facility, Anthropological Studies Center, Sonoma State University.

P&SRR Co. (Petaluma and Santa Rosa Railway Company). 1907. [Railway Commission Map 'A'] Map of final location Petaluma & Santa Rosa Ry. Co., Petaluma to Forestville, Sebastopol to Santa Rosa, Sonoma Co., Cal. Sheet 26 of 33. *Courtesy of Sonoma County History & Genealogy Library.*

Pacific Rural Press. 1874. Agricultural notes. October 17. *Courtesy of California Digital Newspaper Collection.*

Pacific Rural Press. 1876. An important lease. September 30. *Courtesy of California Digital Newspaper Collection.*

Pacific Rural Press. 1880a. Agricultural notes. April 17. *Courtesy of California Digital Newspaper Collection.*

Pacific Rural Press. 1880b. Extraordinary yield of hay. November 6. *Courtesy of California Digital Newspaper Collection.*

Pacific Rural Press. 1882. Agricultural notes. July 15. *Courtesy of California Digital Newspaper Collection.*

Petaluma Journal & Argus. 1869. Sonoma County items. November 27. *Courtesy of Newspapers.com.*

Petaluma Weekly Argus. 1881. Sonoma County items. January 7. *Courtesy of Newspapers.com.*

PWA (Philip Williams & Associates LP). 2004a. *Laguna de Santa Rosa feasibility study: Year one geomorphic investigation: Final report.*

PWA (Philip Williams & Associates LP). 2004b. *Sediment sources, rate and fate in the Laguna de Santa Rosa, Sonoma County, CA: Final Report, Volume 2.*

Priestly HI. 1946. *Franciscan explorations in California*. Glendale, CA: The Arthur H. Clark Company.

Ramsar (The Ramsar Convention on Wetlands). 2017. The list of wetlands of international importance, 2 February 2017. http://www.ramsar.org/sites/default/files/documents/library/sitelist.pdf. Accessed March 13, 2017.

Rhemtulla JM, Mladenoff DJ. 2007. Why history matters in landscape ecology. *Landscape Ecology* 22(1-3).

Richardson CJ, Qian SS. 1999. Long-term phosphorus assimilative capacity in freshwater wetlands: a new paradigm for sustaining ecosystem structure and function. *Environmental Science & Technology* 33(10): 1545-1551.

Ross SAT. 1914. *Recollections of a pioneer: Green Valley, Sonoma Co., Calif.* BANC MSS C-D 5152. *Courtesy of The Bancroft Library, UC Berkeley.*

Santa Rosa Plain Conservation Strategy Team. 2005. *Santa Rosa Plain conservation strategy.* https://www.fws.gov/sacramento/es/recovery-planning/Santa-Rosa/es_recovery_santa-rosa-strategy.htm. Accessed March 20, 2017.

Santa Rosa Soil Conservation District, Sonoma County Flood Control and Water Conservation District (SCFCWCD), U.S. Department of Agriculture (USDA) Soil Conservation Service. 1958. *Watershed work plan: Central Sonoma Watershed,* Sonoma County, California. https://nrm.dfg.ca.gov/FileHandler.ashx?DocumentID=19966. Accessed March 20, 2017.

Sanyal SK, De Datta SK. 1991. Chemistry of phosphorus transformations in soil. In *Advances in Soil Science, Volume 16*, ed. B.A. Stewart. New York: Springer-Verlag.

SCFCWCD (Sonoma County Flood Control and Water Conservation District). 1965. *Flood!! December 1964-January 1965. Courtesy of Sonoma State University Library.*

Searls C. 1897. Mrs. M.A. Peterson, Respondent, v. City of Santa Rosa, Appellant. In *Report of cases determined in the Supreme Court of the State of California, C. P. Pomeroy, Reporter, Volume 119.* San Francisco, CA: Bancroft-Whitney Company. Gaye LeBaron Collection, Box 043. *Courtesy of Sonoma State University Library Special Collections.*

The Sebastopol Times. 1903a. Metropolis of the thrifty Gold Ridge - Sebastopol making vigorous growth. January 2. *Courtesy of Sonoma State University Library.*

The Sebastopol Times. 1903b. The best part of the county. January 2. *Courtesy of Sonoma State University Library.*

SFEI-ASC (San Francisco Estuary Institute-Aquatic Science Center). 2014. NCARI: North coast aquatic resource inventory mapping.

Shelton AC. 1911. Nesting of the California Cuckoo. *The Condor* 13(1):19-22.

Sloop C, Honton J, Creager C, et al. 2007. *The altered Laguna: a conceptual model for watershed stewardship.* Santa Rosa, CA: Laguna de Santa Rosa Foundation.

Sonoma County Democrat. 1861. Grizzly killed. May 28. *Courtesy of Sonoma State University Library.*

Sonoma County Democrat. 1863. County official press. February 28. *Courtesy of California Digital Newspaper Collection.*

Sonoma Democrat. 1879. Catfish. June 7. *Courtesy of Sonoma State University Library.*

Swetnam TW, Allen CD, Betancourt JL. 1999. Applied historical ecology: Using the past to manage for the future. *Ecological Applications* 9(4): 1189-1206.

Tancreto R, Rivera L. 1987. *Santa Rosa chronolgy (memorandum).* Santa Rosa, CA: California Regional Water Quality Control Board, North Coast Region.

Taylor B. 1862. *At home and abroad: a sketch-book of life, scenery and men.* New York: G.P. Putnam.

TetraTech. 2015a. *Laguna de Santa Rosa nutrient analysis (revised).* Prepared for U.S. EPA Region 9 and North Coast Regional Water Quality Control Board.

TetraTech. 2015b. *Laguna de Santa Rosa sediment budget.* Prepared for U.S. EPA Region 9 and North Coast Regional Water Quality Control Board.

Thompson RA. 1877a. *Historical and descriptive sketch of Sonoma County, California.* Philadelphia: L. H. Everts & Co.

Thompson TH. 1877b. Map number six (Russian River, Santa Rosa, Analy Townships) in Historical atlas map of Sonoma County, California. Compiled, drawn and published from personal examinations and actual surveys. Oakland, CA: Thos H. Thompson & Co. *Courtesy of David Rumsey Map Collection.*

Thompson RA. 1884. *Central Sonoma: a brief description of the township and town of Santa Rosa, Sonoma County, California, its climate and resources.* San Francisco: W. M. Hinton & Co.

Tracy CC. 1859a. *Field notes of the obsolete survey of the Rancho Llano de Santa Rosa, Joaquin Carillo, confirmee.* Book G10. U.S. Department of the Interior, Bureau of Land Management. *Courtesy of Bureau of Land Management.*

Tracy CC. 1859b. Plat of the Llano de Santa Rosa, finally confirmed to Joaquin Carillo. Surveyed under instructions from the U.S. Surveyor General. Land Case Map E-131. *Courtesy of The Bancroft Library, UC Berkeley.*

USDA (U.S. Department of Agriculture). 1942. [Aerial photos of Sonoma County] Flight COF-1942. U.S. Department of Agriculture, Soil Conservation Service.

USDA (U.S. Department of Agriculture). 2009. *EvegTile22A_07_24k_v1.* USDA Forest Service, Pacific Southwest Region, Remote Sensing Lab. McClellan, CA.

USDC (U.S. District Court California, Northern District). ca. 1840a. [Diseño del Rancho El Molino o Rio Ayoska : Sonoma Co., Calif.] Land Case Map B-492. *Courtesy of The Bancroft Library, UC Berkeley.*

USDC (U.S. District Court California, Northern District). ca. 1840b. [Diseño del Rancho Llano de Santa Rosa : Calif.] Land Case Map B-128. *Courtesy of The Bancroft Library, UC Berkeley.*

USGS (U.S. Geological Survey). [1933-5]1942. Sebastopol Quadrangle, California: 15-Minute series (Topographic). 1:62,500.

USGS (U.S. Geological Survey). 1913. *Topographic instructions of the United States Geological Survey.* Department of the Interior, United States Geological Survey. Washington: Government Printing Office.

U.S. Surveyor General's Office. 1866. Map of Township No. 7 North, Range No. 9 West (Mount Diablo Meridian). San Francisco, CA. *Courtesy of Bureau of Land Management.*

Uusi-Kämppä J, Turtola E, Hartikainen H, Yläranta T. 2001. The interactions of buffer zones and phosphorus runoff. In *Buffer Zones: Their Processes and Potential in Water Protection, The Proceedings of the International Conference on Buffer Zones, September 1996*, ed. Nick Haycock, Tim Burt, Keith Goulding, and Gilles Pinay. Haycock Associated Limited.

Vallejo MG, Farris G, Beebe R. 2000. *Report of a Visit to Fort Ross and Bodega Bay*. California Mission Studies Association, Occassional Paper #4. http://www.fortross.org/lib/46/report-of-a-visit-to-fort-ross-and-bodega-bay-in-april-1833-by-mariano-g-vallejo.pdf. Accessed March 13, 2017. *Courtesy of Fort Ross Conservancy Library.*

Verhoeven JTA, Setter TL. 2010. Agricultural use of wetlands: opportunities and limitations. *Annals of Botany* 105(155-163).

Waaland M. 1989. *Long-term detailed wastewater reclamation studies: Santa Rosa subregional water reclamation system technical memorandum no. W1: Baseline evaluation of Laguna de Santa Rosa wetlands and natural resources.* Golden Bear Biostudies.

Waaland M. 1990. History of human use and modification of the Laguna ecosystem. In *History, land uses and natural resources of the Laguna de Santa Rosa*, ed. David W. Smith Consulting, 6-1 to 6-10. *Courtesy of Sonoma County History & Genealogy Library.*

Walker WS. 1880. *Glimpses of hungryland; or, California sketches. Comprising sentimental and humorous sketches, poems, etc., a journey to California and back again, by land and water; incidents of every-day life on the Pacific coast, why I came, what I saw, and how I like it.* Cloverdale: Reveille Publishing House.

Watson EB, Dean D, Zinn CJ, et al. 1915. Soil map, California. Healdsburg sheet. U.S. Department of Agriculture, Bureau of Soils. *Courtesy of Alabama Maps, University of Alabama.*

Watson EB, Dean WC, Zinn CJ, et al. 1917. *Soil survey of the Healdsburg area, California.* Advance sheets, field operations of the Bureau of Soils, 1915. U.S. Department of Agriculture, Bureau of Soils. *Courtesy of Bureau of Land Management.*

Whitacre TH. 1853. *Copy of field notes of survey of Township 6, 7, 8, 9, and 10 North of Ranges numbers 6, 7, 8, 9, and 10 West of Mount Diablo Meridian in the State of California.* Book 210-5. U.S. Department of the Interior, Bureau of Land Management. *Courtesy of Bureau of Land Management.*

Winzler & Kelly-GHD. 2012. *Laguna-Mark West Creek watershed planning scoping study: Screening technical memorandum.*

Wright JP, Jones CG, Flecker AS. 2002. *An ecosystem engineer, the beaver, increases species richness at the landscape scale.* Oecologia 132:96-101.

WSI (Watershed Sciences, Inc). 2013. [3-ft, LiDAR-derived bare earth digital elevation model (DEM) for Sonoma County.] NASA Grant NNX13AP69G, University of Maryland, and the Sonoma Vegetation Mapping and LiDAR Program. Data available at http://sonomavegmap.org/.

APPENDIX: SPECIES

Common Name	Scientific Name
Beaver	*Castor canadensis*
Boxelder	*Acer negundo*
Bulrush	*Scirpus* and *Bolboschoenus* spp.
California freshwater shrimp	*Syncaris pacifica*
California tiger salamander	*Ambystoma californiense*
Cattail	*Typha* spp.
Coho salmon	*Oncorhynchus kisutch*
Deer	*Odocoileus hemionus*
Elk	*Cervus elaphus*
Grizzly bear	*Ursus arctos horribilis*
Ludwigia	*Ludwigia* sp.
Oak	*Quercus* spp.
Black oak	*Quercus kelloggii*
Valley oak	*Quercus lobata*
Oregon ash	*Fraxinus latifolia*
Pronghorn antelope	*Antilocapra americana*
Steelhead	*Oncorhynchus mykiss*
Tule	*Schoenoplectus* spp.
Western yellow-billed cuckoo	*Coccyzus americanus*
Willow	*Salix* spp.

Photo by Sean Baumgarten,
January 13, 2017

www.ingramcontent.com/pod-product-compliance
Lightning Source LLC
Chambersburg PA
CBHW041731210326
41598CB00008B/842